The Birth of the Cell

The Birth of the Cell

Henry Harris

Yale University Press
New Haven and London

First published in paperback 2000

Set in Simoncini Garamond by Servis Filmsetting Ltd, Manchester
Printed in Great Britain by St Edmundsbury Press

Library of Congress Cataloguing-in-Publication Data

Harris, Henry, 1925–
The birth of the cell/Henry Harris.
Includes bibliographical references and index.
ISBN 0–300–07384–4 (hbk.)
ISBN 0–300–08295–9 (pbk.)
1. Cytology—History. I. Title.
QH577.H37 1998
571.6′09—dc21 98–18623
CIP

A catalogue record for this book is available from the British Library.

2 4 6 8 10 9 7 5 3

For A. F. H.
Fifty years on

Give me an organic vesicle endowed with life and I will give you back the whole of the organized world.

(Raspail)

Contents

Illustrations

Preface

There can be very few ideas in the history of biology that have had greater influence than what is known as the cell doctrine: that the bodies of all forms of life are composed of independent, but cooperative, units that we now call cells. How the evidence for this doctrine was assembled is the subject of this book. It is largely based on a reading of the original texts. I have been reading them with pleasure for the best part of half a century, with pleasure because, before the advent of the multi-authored twentieth century, a scientific paper usually afforded a glimpse, and sometimes a vivid picture, of the author. For this reason, and to ensure authenticity, I have added an appendix in which all verbatim extracts are given in the languages in which they were written; I am not so skilful a translator that I can reliably capture the nuances and overtones of the original in another tongue, if this can be done at all. In the body of the text (indicated by asterisks) I have provided literal translations into English of all the foreign quotations. Any errors or infelicities that these translations might contain are my own. Close adherence to the original texts has, however, enabled me not only to minimize the distortion that appears to be an inevitable accompaniment of writing history, but also to deal with the main protagonists as personalities and not merely as ciphers engaged in the delivery of information. This cannot, of course, be done without casting them in their political and social background, especially in the nineteenth century when European rivalries often coloured the content and style of scientific communications. Analysis of a scientific movement, although necessarily concerned in the main with empirical observations, both accurate and inaccurate, is also a fragment of social history.

With the passage of the years, my reading of the biological literature of earlier times has left me dissatisfied with many standard accounts of the origin of the cell doctrine, and in particular the perfunctory versions given in general textbooks. More often than not, what is said has been determined by received wisdom rather than meticulous historiography. I have therefore, for the most part, eschewed the secondary sources; but I have made room for a good deal of biographical material, partly because this is of interest in its own right, and partly because what some investigators wrote, how they wrote it and how it was received was to some extent determined by the nature of their lives.

The history of the cell doctrine, as it is commonly expounded, is dominated by the names of Schleiden, Schwann and Virchow, but if one is more interested in those who made the discovery rather than those who elaborated on it, then it seems to me that, among others, Dumortier, Purkyně, Valentin and Remak deserve a more generous treatment than most subsequent historians have, in fact, given them. My aim has been to present a less schematic version of events and to show how, out of a sea of error and confusion, an approximation to the truth finally emerged.

I thank Professor Herman Waldmann of the Sir William Dunn School of Pathology, University of Oxford, for his generous support of this work, and Mrs Valerie Boasten for her help in the preparation of the typescript. Dr Robert Baldock of Yale University Press greatly enlivened the text by his comments and suggestions.

Henry Harris
Oxford
Hilary, 1998

CHAPTER 1

The Early Microscopists

Most accounts of the origin of the cell doctrine begin with the microscopic observations of the polymathic Robert Hooke (1635–1703), who was appointed curator of experiments for the Royal Society in 1662 and whose *Micrographia* appeared in 1665. The development of the subject is usually presented as a discontinuous process determined by successive improvements in the construction and use of the microscope. But although Hooke's celebrated drawing of a thin section through a fragment of cork undoubtedly showed the thickened walls of dead cork cells, Hooke did not for a moment suppose that these structures were the residual skeletons of the basic subunits of which all plants and animals were constituted. Nor would he necessarily have imagined, if he had thought of basic subunits at all, that they would have the size and shape of the cork cavities that he had observed or that they would have an essentially uniform structure. A hundred and fifty years later, in the early part of the nineteenth century, the many microscopists who scoured animal tissues in search of their basic subunits, and who found themselves hopelessly ensnared in a panoply of spherical optical artefacts, were convinced that uniform basic subunits must exist and were confident that they would soon be found. This change in outlook was not driven by advances in microscopy; it was a reflection of the radical reorientation that had taken place in the assumptions that scholars made about the nature of the universe and the matter of which it was composed.

To set this reorientation in its appropriate context we must revert to a debate that exercised the minds of philosophers in Greece in the third and fourth centuries BC. Regrettably, our knowledge of the writings of Leucippus and Democritus, the founding fathers of Greek atomism, is limited largely to hostile accounts provided by their opponents, in particular Aristotle and Simplicius; so we cannot be at all sure of the details of the position that the original atomists advocated.[1] But it is clear that their model of the universe had at its centre the idea that all matter was ultimately composed of indivisible subunits (atoms) which differed from each other, not in their composition, but only in their size and shape; and the only change that they were permitted to undergo was a change in position. This 'corpuscularian hypothesis' as Barnes[2] terms it, was dismissed by Aristotle as logically incoherent. Aristotle argued that all matter was seamlessly continuous and that

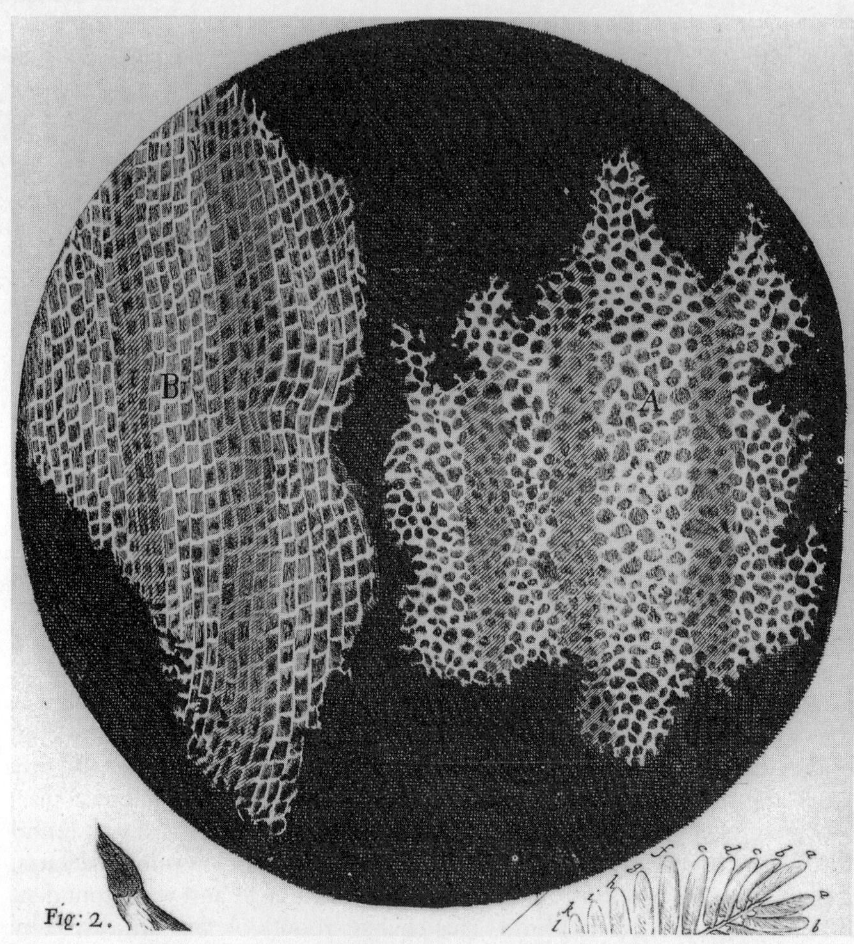

1. Cross-section (A) and longitudinal section (B) of thin sections of cork

discontinuities in which there was no matter could not exist. Although the atomism of Democritus was accepted and further elaborated by Epicurus[3] and received honourable mention in the *De rerum natura* of Lucretius, the views of Aristotle decisively prevailed. We find no exposition or discussion of atomism in the literature of Europe in the Middle Ages. It is likely that the revival of atomism had its origin in the Renaissance with the recovery of *De rerum natura*, but the Epicurean position adopted by Lucretius found few followers much before the early part of the seventeenth century. The first formal exposition of a molecular model of matter in modern times appears to have been that of Isaac Beeckman in 1620.[4] Galileo's *Il Saggiatore*, which was published in 1623, shows that he was already a committed atomist by that time; and this was also true of Marco Aurelio Severino whose *De recondita abscessuum natura* (Concerning the hidden nature of abscesses) of 1632 was probably the first treatise on surgical pathology, and of Marcello Malpighi whose discovery of capillaries in 1661 inaugurated the science of histology. Pierre Gassendi's influential philosophical work,[5] which received its definitive formulation in the *Syntagma philosophicum* published posthumously in 1658, was a systematic attempt to displace the seamless universe of Aristotle with a Catholicized version of Epicurus. Newton was aware of the work of Gassendi, and in his *Quaestiones*, probably initiated in 1664, he already shows a preference for an atomistic view of the universe rather than the uninterrupted continuity advocated by Descartes. A celebrated paragraph in the *Optics*, published in 1704, presents the outline of an atomism that is not too different from that attributed to Democritus. For Gassendi, a churchman, there was, however, the formidable difficulty of stretching Epicurean materialism to accommodate the soul. Gassendi offered no solution to this problem other than the scarcely explanatory proposal that the soul in some way governed the activities of the atoms that constituted living forms. Leibniz proposed a more radical solution. In his *Monadologie* of 1714 he abandoned altogether the homogeneous atoms of Democritus and substituted a hierarchy of monads with various properties including, at the apex of the system, monads that are not only self-generating, but that also have souls. The contemporary reaction to Leibniz's ideas makes it clear that by the early part of the eighteenth century, despite the continued adherence of orthodox theologians to essentially Aristotelian doctrines, most experimentalists and many philosophers were at ease with the general proposition that the universe was basically atomistic; and, for the most part, arguments centred not on whether such basic subunits existed but on what properties they had and how their behaviour might explain the phenomena of the observable world. When microscopists began to look at the tissues of living forms they already had in their minds a view of matter as an aggregate of more or less uniform microscopic components. It is therefore understandable that when they saw everywhere agglomerations of more or less spherical halations, they concluded that these optical illusions were the fundamental subunits of animate matter; and when they actually saw cells they had no idea what they were.

Although the development of the modern microscope has its roots in the work of the seventeenth-century Dutch lens-grinders, and in particular the construction of a microscopical device by Cornelis Drebbel somewhere in the vicinity of 1620, it seems likely that the microscopic examination of living tissues, albeit with inferior lenses, was initiated in Italy. In the *Sidereus nuncius*, which was published in 1610, Galileo described the use of the telescope, but, as pointed out by Belloni, the editor of the *Opere scelte di Malpighi*,[6] Galileo had by 1614 adapted lenses for use in microscopy and had described the appearance of the cuticle of the fly, which he found 'interamente coperto di pele' (completely covered in fur). Nicolas-Claude Fabri de Peiresc published observations on the fine structure of mites in 1622, and in 1624 Cesi, the founder of the Academy of the Lincei described 'un occhialino

2. Pierre Gassendi (1592–1655)

per veder da vicino le cose minime' (a lens for looking closely at the smallest of objects); it was the Lincei who gave it the name 'microscopio'. The *Melissographia Lincea* of Francesco Stelluti, which presented his microscopical observations on bees, appeared in 1625. To begin with, the microscope was used to examine the surfaces of objects, and in particular insects, but in the work of Giovanni Battista Odierna we find microscopy combined with microdissection. His *L'occhio della mosca*, which appeared in 1644, presents a detailed description of the internal structure of the eye of a fly. All this long antedates Hooke's *Micrographia* (1665).

Hooke (no likeness of him appears to have survived) began to develop his

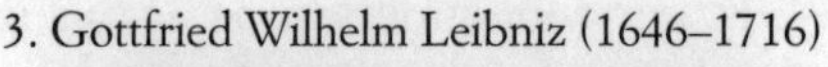

3. Gottfried Wilhelm Leibniz (1646–1716)

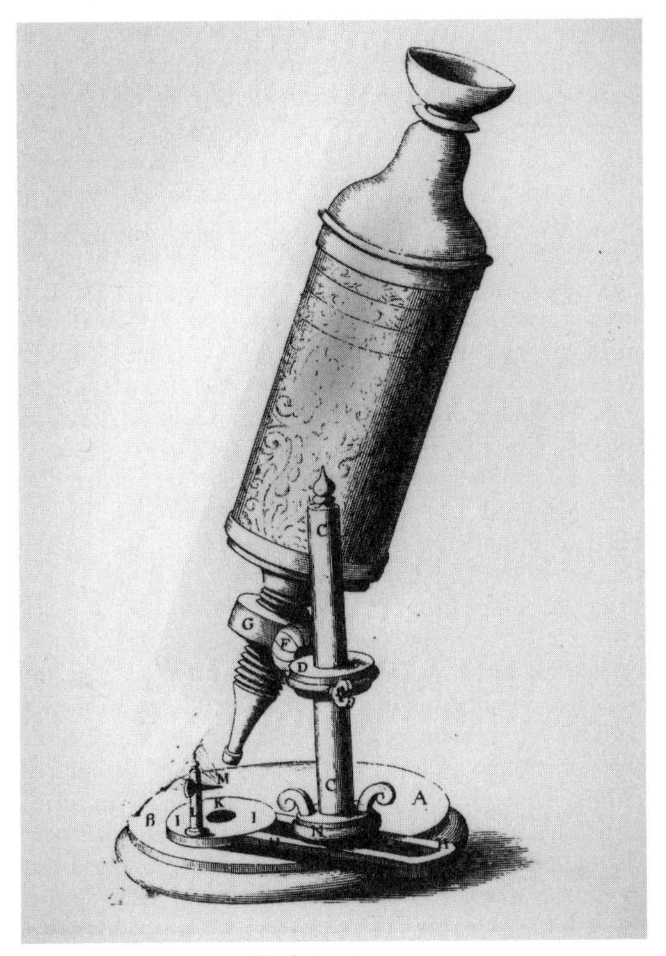

4. Hooke's microscope

compound microscope in about 1660. The final version illustrated in the *Micrographia* contains only two lenses, a small plano-convex object lens with the convexity facing the object and a larger plano-convex ocular with the convexity facing the eye. The tube connecting the two lenses was filled with water to enhance clarity. Hooke experimented with a third lens placed between the object and ocular lenses, but found that this produced too much light scattering to be useful. In view of Hooke's apparent primacy in the discovery of cells, it is important to establish both what he saw and what he did not see. His first microscopical observations on plant material were made on petrified and burnt wood or on charcoal, in which, with a hand lens, he found numerous cavities or 'pores'. He communicated his findings to Evelyn who referred to them in his *Sylva* of 1664. With a microscope, however, he saw, in addition to the larger pores visible with a hand lens, a tightly apposed array of much smaller cavities. It is not clear whether these were also vessels or whether they represented the petrified or carbonized walls of cells. In any case, Hooke believed that in the living state they were 'filled with the natural or innate juices of those Vegetables'. There is no doubt, however, that in the microscopical examinations that he made of thin sections of cork, cut with a penknife 'sharpened as keen as a razor', he observed cell walls, and in each of the empty spaces enclosed by these walls, there must once have been at least one living cell. It is possible that the walled cavities that he saw in cork were not the same as those that he had previously seen in petrified or burnt wood, for of the cavities in cork he says that they 'were indeed the first *microscopical* pores I ever saw, and perhaps, that were ever seen, for I had not met with any Writer or Person that had made any mention of them before this'. Given that one of the common meanings of the Latin word *cella* was a small room or cubicle, Hooke's use of the word 'cell' to describe these small enclosed cavities was apposite.

It has been suggested that Hooke also saw living cells. This, however, is doubtful. He was aware that the 'cells' that he saw in cork contained air, and he uses this fact to explain the properties of cork – its light weight, its compressibility, its ability to float in water. But he also thought that in the living state these 'cells' were 'the channels or pipes through which the *Succus nutritius*, or natural juices of Vegetables are conveyed'. He regarded the transverse walls as diaphragms that divided the channels into sections, and looked in vain for pores that might permit sap to pass from one cell space to the next. He tried, again without success, to force fluid through the network of contiguous spaces. He thought that his failure to detect a system of communication that could permit the movement of fluid was due to the still modest resolving power of his microscope, and he thought it probable that the problem would be resolved in due course by the development of better microscopes. It is obvious that Hooke made no distinction between the cellular spaces that he saw in cork and the vascular system of plants. Indeed, he thought these spaces were analogous to the 'vessels in sensible creatures', and he searched for a system of valves. This confusion between vessels containing sap and cells containing a semi-fluid cytoplasm persisted until the nineteenth century when Dujardin (1835), Purkyně (1839) and von Mohl (1846)[7] finally delineated and proposed a nomenclature for the contents of the cell. It is of interest that, after calculating the size of the cellular cavities in cork, Hooke was surprised that channels so small could carry all the nutrients required by a large plant, for these spaces were 'yet so exceeding small, that the *Atoms* which Epicurus fancy'd would go neer to prove too bigg to enter them much more to constitute the fluid body in them'. Evidently, in the seventeenth century, the ideas that experimentalists had about the size of Epicurean atoms were still very vague.

Hooke claims to have seen similar 'cells' in the soft tissues of a number of other plants: 'the pith of an Elder, or almost any other Tree, the inner pulp or pith of the Cany, hollow stalks of several other Vegetables: as of Fennel,

Carrets, Daucus, Bur-docks, Teasels, Fearn, some kinds of Reeds, &c'. It seems possible that what Hooke observed in these situations were cross-sections of vascular bundles of varying dimensions, sometimes containing sap. A honeycomb of grossly thickened cell walls enclosing empty spaces is not what the microscope reveals in the soft tissues of the other plants that Hooke studied, and, regrettably, he gives no illustrations of the cellular structures that he saw in these plants. There is, however, one drawing in the *Micrographia* that has been adduced as evidence that Hooke did see living cells. Charles Singer in his well-known *Short History of Biology* (1931) points out that in Hooke's representations of the under-surface of the leaf of a stinging nettle,[8] the outlines of cells can be seen. This is a misconception. The epidermal cells of nettle leaves are not flat polygons as indicated by the outlines noted by Singer. As in other dicotyledonous plants, these cells have sinuous anticlinal walls, and there are in fact between 100 and 200 such cells in each of the polygons shown in Hooke's drawing. The outlines of the polygons are not cell walls but a reticulum of small vascular bundles.[9] Hooke also studied sections of feathers where he again found walled spaces, not ordered in linear arrays as in cork, but agglomerated in an apparently haphazard fashion resembling 'solid or hardened froth or a *congeries* of small bubbles'. Hooke offers no suggestions for the functions of these walled spaces in feathers, but fails, as in cork, to find a system of communication between them. My reading of the relevant passages of the *Micrographia* leads me to conclude that Hooke did see the walls of individual cells in cork, but he misunderstood their function and he clearly had no conception of what, in the living state, occupied the spaces within these walls. There is, in my view, no convincing evidence that he saw individual living cells in any other tissue.

Six years after the appearance of Hooke's *Micrographia*, the Royal Society, which by then had already acquired a European reputation, received two manuscripts that proved to be the foundation stones of our modern understanding of the fine structure of plants. One of these manuscripts came from Nehemiah Grew (1641–1712), the other from Marcello Malpighi (1628–94). Although there is no question but that the two authors made their discoveries independently of each other, the exiguous margin of priority that separated the receipt of the two communications has been the subject of endless controversy. This controversy was fuelled by some mischief generated by Schleiden in his *Grundzüge der wissenschaftlichen Botanik* (Principles of scientific botany) (1842). Schleiden asserts, for reasons that are difficult to understand unless they are simply an expression of anglophobia, that Grew in his capacity as Secretary of the Royal Society held up the publication of Malpighi's work in order to ensure his own priority. As far as can be ascertained, there is nothing in this accusation.

To begin with, Grew was not appointed Secretary to the Society until 1677, whereas the two manuscripts arrived in December 1671. On present evidence, which is unlikely to be substantially modified, Grew first sent part of his draft to Francis Glisson, Regius Professor of Physic at Cambridge and

5. Nehemiah Grew (1641–1712)

a Foundation Fellow of the Royal Society, who sent it on to Henry Oldenburg, one of the Society's first two secretaries. Oldenburg sought the opinion of John Wilkins, the other Secretary, who suggested that Grew send the whole of the draft to the Society. The Society's Council ordered the printing of the manuscript on 11 May 1671. Although dated 1672, the work appeared on 7 December 1671. Grew wrote the *Anatomia vegetalium inchoata* initially in Latin, but the first edition of the book bears the English title *The Anatomy of Vegetables Begun* and was published by and for the Royal Society. The second edition was translated into French in 1675 and into Latin in 1678. This second edition forms the first part of the collective

6. Marcello Malpighi (1628–94)

Anatomy of Plants that summarizes Grew's work up to 1682, when the volume appeared.

Between 1672 and 1682 Grew published a number of papers and books dealing with various aspects of plant microanatomy, and in considering his contribution to the subject it is important to remember that in some respects his views changed over this decade. He first called the cell spaces 'bubbles' and thought that they were formed by a process akin to fermentation, analogous perhaps to the bubbles formed in bread when it is baked. In 1672 he referred to the cellular spaces as 'pores', thought some of them were divided by partitions and noted that there were still smaller pores at the sides of the larger ones. Later, he adopted Hooke's term 'cells' or Malpighi's notion of 'bladders'. In the *Anatomia radicum* he says: 'The microscope reveals that the substance of the bast is nothing other than a huge mass of small cells or rigid bubbles.'* Nonetheless, even in 1682, he still believed that these bladders, when first formed, were full of air: when the seed was penetrated by the sap, he thought it formed a coagulum that was transformed by a process of fermentation 'into a *Congeries of bladders*: for such is the *Parenchyma* of the whole *Seed*'. Grew invented the term 'parenchyma' to describe this fenestrated soft tissue, and although the cellular spaces were sometimes full of what he thought to be sap, the idea that they were fundamentally air spaces persisted. The model that Grew proposed in 1682 confirms this. He likened the parenchyma to bone-lace, where the walls of the fenestrations were 'threads or fibres' held down by the vessels like pins on a cushion. Since he thought of the parenchyma in three-dimensional terms, he envisaged many such lace-like layers making a series of closed cavities: 'And this is the true texture of a *Plant*, and the *general composure*, not only of a *Branch*, but of all other *Parts* from the *Seed* to the *Seed*'.

In the preface of *The Anatomy of Plants* Grew makes the following statement: 'Not long afterwards I received news from London that the same day in which my book was presented there was also presented a manuscript (without figures) from Signor Malpighi upon the same subject, dated Bologna November 1st 1671.' This is, in essence, a priority claim on Grew's part, or, at the very least, a claim for independent work carried out prior to the publication of his book, and hence prior to the receipt of Malpighi's manuscript. I do not think it would ever have been in doubt that both men had worked on plant materials for some time before they sent their abstracts to the Royal Society, but it may well have been the statement in the preface of *The Anatomy of Plants* that provoked Schleiden's suspicious remark that the officers of the Royal Society connived at the rapid publication of Grew's work in order to establish his priority over Malpighi. It is the case that the material submitted by Grew was printed at once, and the published work, bearing the date 1672, was presented to the Society even earlier, in December

* *Note*: Asterisks throughout the text denote translations of extracts given in the original in the Appendix (pp. 185–200).

1671; whereas the first part of Malpighi's work did not appear in print until 1675. However, Malpighi's initial abstract, written in elegant Latin, contained only a fraction of the material that eventually appeared as his *Anatomes plantarum* (1675 and 1679).

The abstract, dated November 1671, was entitled simply *Anatomes plantarum idea* and forms the first part of the *Anatomes plantarum pars prima* which appeared in 1675. This closes with an *Appendix repetitas auctasque de ovo incubato observationes continens* (Appendix containing repeated and extended observations on the incubation of the egg) and is dated October 1672. The work that appeared in 1679 was the second volume of Malpighi's studies and bore the title *Anatomes plantarum pars altera*. It is not, at this remove, clear why Grew's work was published virtually at once, whereas there was a delay of some three or four years before Malpighi's work first appeared in print. One possible reason is that Malpighi's *Anatomes plantarum pars prima*, dated 1671 and 1672, contains much more than the *Anatomes plantarum idea* that formed the original abstract; but whether the delay was due to Malpighi or the Royal Society is at present obscure. Another possible reason is that the *Anatomes plantarum pars prima* contains illustrations that did not appear in the *Anatomes plantarum idea*; and Grew, in 1682, does not fail to mention this. Nonetheless, it is difficult to see that the absence of illustrations would have justified a delay of three or more years, unless Malpighi was very slow in delivering them. In any case, when Grew first received Malpighi's *Anatomes plantarum idea*, he was so impressed with the depth of what he read that he proposed to give up work on plant microanatomy altogether; but it appears that the Royal Society prevailed on both men to continue their investigations.

Given that both Malpighi and Grew had worked independently on plant material for some time before submission of their manuscripts to the Royal Society, and given that the two submissions were more or less simultaneous, the controversy about priority is of interest mainly because it illustrates the competitiveness of scientists even in the seventeenth century. There is, perhaps, more point in examining the overall contributions of the two men. In this connection, two things should be borne in mind. First, Grew's major work *The Anatomy of Plants* of 1682 embodied not only his own work, but also ideas that he had gleaned from Malpighi, whereas Malpighi's two publications of 1675 and 1679 deal only with his own work. And second, Grew's work, although he was an MD, was almost entirely botanical, whereas the microanatomy of plants was only a part of the work of Malpighi. The comparison between the two men has usually been made by botanists, from Sachs[10] to Morton,[11] and they have been particularly impressed by Grew's discovery that the vessels of plants were formed by the breakdown of partitions between rows of contiguous cells; by his invention of the word 'parenchyma' to describe the cellular nature of soft tissue; and by the detail shown in his drawings. Naturally enough, Malpighi's work on animal tissues is largely ignored by those whose interest is centred on plants. But a glance at a medical diction-

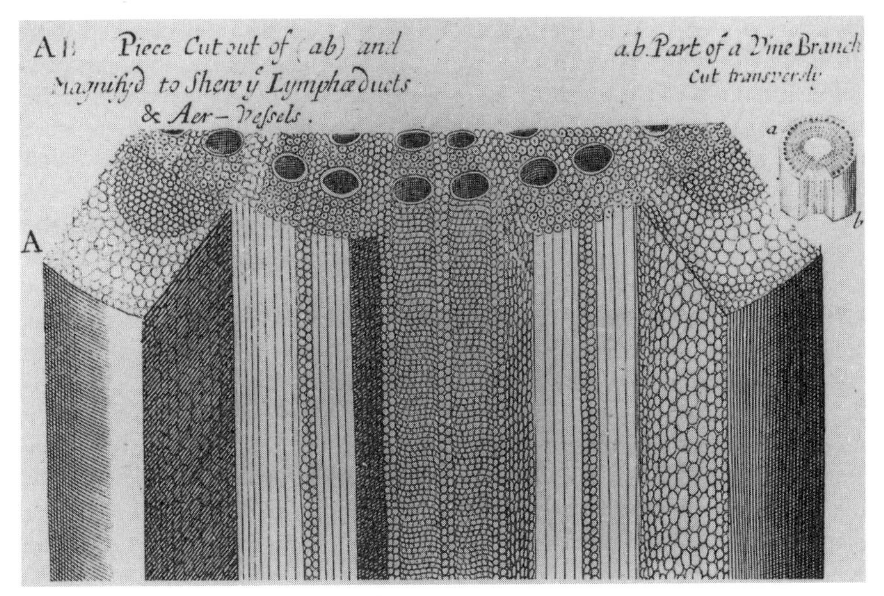

7. Grew's illustration of sections through a vine stem

ary reveals that there are, among other Malpighian eponyms, the Malpighian bodies, Malpighian cells, Malpighian corpuscles, Malpighian vesicles, Malpighian pyramids, Malpighian stigmata, the Malpighian stratum and the Malpighian *rete mirabilis*, all of which testify to Malpighi's primary role in the microscopic analysis of the animal body. In fact, Malpighi can be regarded as one of the founding fathers, if not the founding father, of microscopic zootomy as well as phytotomy.

Malpighi was thirteen years older than Grew at the time of the first submission of their manuscripts to the Royal Society and, as professor of medicine at Bologna, was already a scientific figure of European reputation. It is likely that he and his colleagues initiated the histological study of animal tissues some years before Grew began his work on the structure of plants. Grew was, after all, only thirty years old when he sent his manuscript to Glisson. Malpighi, with his close colleague Giovanni Alfonso Borelli (1608–79), a professor at Pisa, and with Claudius Auberius (Claude Aubery or Aubry) began a systematic study of animal (and possibly plant) tissues at least a decade before that.[12] In the *De motu animalium*, published in Rome in 1681, Borelli describes the spiral fibres in cardiac muscle and reports the discovery as having been made in 1657. Malpighi claimed that the discovery was his. Be that as it may, in two letters written to Borelli in 1661, he did describe for the first time the capillary bed in the lungs of frogs. In these two letters, published as *Sui pulmoni* (*De pulmonibus*), he illustrates the use of the microscope at various magnifications with both transmitted and reflected light, and a range of anatomical procedures including drying, dissecting, insufflation and vascular perfusion. What is more, he often worked with tissues in the living state. His discovery of the capillary bed at long last provided a solution to the outstanding problem posed by Harvey's model of the circulation of the blood. Having rejected Galen's postulate that there were invisible pores in the interventricular septum of the heart, and having failed to find the hypothetical peripheral anastomoses proposed by Galen, Harvey was left with no plausible explanation for the passage of blood from the venous to the arterial system. Malpighi's discovery of pulmonary capillaries provided that explanation.

Because of his extensive studies on animal tissues, Malpighi's observations on plant tissues were much influenced by the analogies between the two systems. Thus, he called the vessels in the xylem (the plant tissue that conducts water) tracheae, which he thought were analogous to the respiratory tubes of insects. He even imagined that he saw peristalsis (involuntary contractions) in them. Like Grew, he believed that they were filled with air. It is likely that, eventually, Grew reached the conclusion that the whole of the plant, including the secondary thickening of the stem, was derived from the cellular parenchyma; but in the *Anatomes plantarum pars prima* of 1675 Malpighi already has illustrations of transverse sections of the stem of *Portulacca*, in which the tissues are entirely composed of cells showing differences in size. Both men studied the formation of the annual rings and noted the differences between spring and autumn wood. The words that Malpighi chose to describe the cellular spaces are of interest in themselves. He did not use Hooke's word *cellulae* but called these spaces by the more sophisticated terms *utriculi* and *sacculae*. In classical Latin *utriculus* means a small bottle, commonly of skin or leather, and *sacculus* is used in two ways to mean either a small bag in general, or a bag for straining liquids. It cannot, of course, be known whether in the seventeenth century these terms still had their classical connotations, but it is clear that in choosing them, rather than *cellulae*, Malpighi had in mind that in the living state these spaces were at some stage filled with liquid and possibly with liquid that was strained through the cell walls. Malpighi would, in all probability, have rejected Grew's analogy with fine bone lace, for he certainly envisaged his *utriculi* or *sacculi* as something more than spaces left vacant by an interwoven three-dimensional network of fibres. In fact, in the *Anatomes plantarum pars prima* Malpighi notes that if the petals of plants are broken up, rows of *utriculi* linked together are released. He actually provides an illustration of such concatenated cells taken from the petal of a tulip.

In the *Anatomes plantarum pars altera* Malpighi discusses parasitism in plants. He describes the formation of a gall produced by the deposition of insect eggs in the plant tissues, and in the *De Gallis* section of the book he gives an accurate illustration of the process. This observation is of general significance because it adds a major piece of evidence against the notion of spontaneous generation. Francisco Redi, personal physician to the Grand Duke of Tuscany, Ferdinand II de' Medici, in the *Esperienze intorno alla gen-*

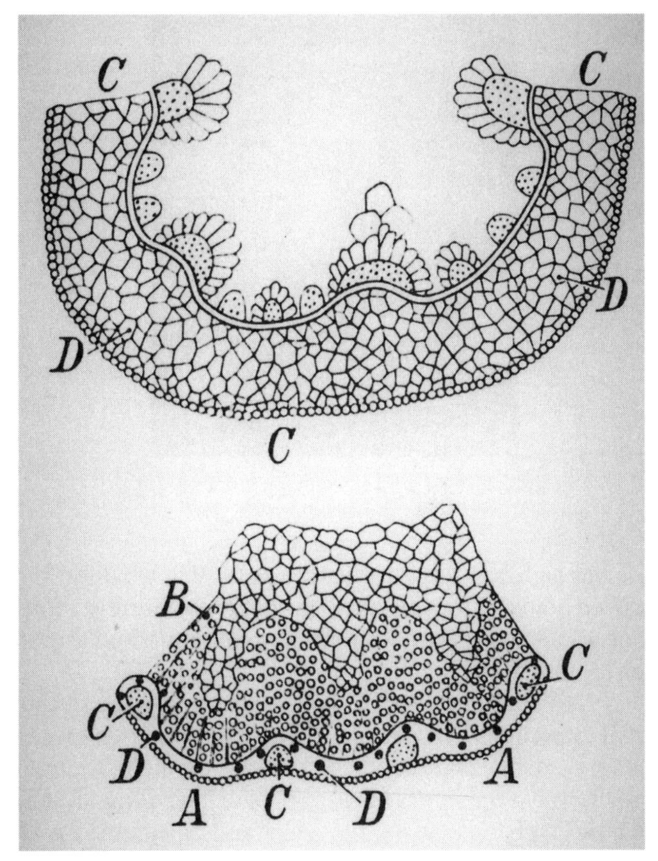

8. Malpighi's illustration of sections through the stalk of *Portulacca*

erazione degl' insetti (Experiments concerning the generation of insects), published in 1668, had already shown that maggots did not arise spontaneously from decaying flesh but were produced from eggs deposited in the meat by flies.

In a study devoted to the rather scholastic question of priority of publication, Pollender[13] reaches the conclusion that in the matter of plant microanatomy the palm should go to Grew. But if one considers the total contribution to science that each of the two men made in the course of a lifetime, one is left with little doubt that Malpighi was the greater figure. Belloni[14] complains that the many contributions of Malpighi have been largely ignored by subsequent historians. As far as historians in the English language are concerned, and especially those interested mainly or exclusively in botanical history, I am inclined to agree with him. It was Malpighi, after all, who initiated the systematic study of microanatomy as a whole and who developed the methodology necessary for this purpose. Howard B. Adelman,

9. Antoni van Leeuwenhoek (1632–1723)

the editor of Malpighi's letters[15] makes the point that whatever rivalry there might have been between the two men, 'nothing but courtesy, mutual respect and generous appreciation of one another's efforts marked their relation in the years that passed till the completion of their work'.

For half a century or more very little was added to the descriptions of plant microanatomy presented by Grew and Malpighi. Leeuwenhoek, who mentions both men by name, confirmed the presence of microscopical cavities in transverse sections of seeds and of the stems of oak seedlings, but his descriptions fall far short of those given by Grew and Malpighi. By 1719, when Leeuwenhoek's *Epistolae physiologicae super compluribus naturae arcanis* (Physiological letters concerning several of nature's mysteries)[16] were published, it appears to have been common knowledge among botanists, as Baker notes in his 1948 review of the 'Cell-Theory',[17] that the tissues of plants, at least, were largely composed of aggregates of small chambers.

CHAPTER 2

Globules, Fibres and Twisted Cylinders

Libraries have been written about the transition from medieval scholasticism to eighteenth-century empiricism, but nowhere is the change in mood better or more succinctly expressed than in the motto of the Royal Society, *Nullius in verba*, a contraction of a phrase from Horace, 'nullius addictus iurare in verba magistri' (not bound to swear by the words of any master),[1] with which most Fellows of the Royal Society would at that time have been familiar. This motto not only asserted the primacy of experiment, as opposed to doctrine, it also rested on an interpretable, and for most purposes mechanical, view of the world. As Johann Andreae, a Protestant divine, put it in his *Reipublicae christianopolitanae descriptio* of 1619: 'If you do not analyse matter through experiment, if you do not improve knowledge through better instruments, you are worthless.'[2] Of course, advances in optical microscopy were fundamental to the progress that led eventually to an understanding of the cell as we now know it, but again it was the assumption that the tangible world was mechanistic and that increased resolution and accuracy of microscopic observation would eventually reveal its secrets that drove the development of better instrumentation.

It appears to have been Johannes Hudde (1628–1704)[3] who first devoted his time to experiments aimed at making better microscopes with simple lenses. He was a celebrated man of his time, at least in the Low Countries, and several Amsterdam investigators dedicated their books to him. It was Hudde who taught both Leeuwenhoek and Swammerdam how to grind good lenses from drops of glass, and it was through the use of microscopes based on such lenses that both men achieved their fame. It remains unclear whether Swammerdam, Leeuwenhoek or Malpighi was the first to see animal cells, the red corpuscles of the blood. The main source of this uncertainty is the fact that Swammerdam's great book, *Biblia naturae*[4] was not published until long after his death. It appeared in 1737 and 1738, whereas Swammerdam lived from 1637 to 1680. Unfortunately the *Biblia naturae* does not accurately date the observations made on human blood. In dissecting the louse *Pediculus*, Swammerdam mentions the effusion of blood from the upper part of the abdomen, but he states that the effusion was composed of globules that resemble those seen in cows' milk. The latter are obviously not red cells, but probably fat globules, and this confusion between the two kinds of particle

10. Jan Swammerdam (1637–80)

persisted for some time in the work of later authors. However, Swammerdam mentions that a few years prior to his observations on the louse, such particles were seen in human blood, where they were slightly reddish and were suspended in a clear fluid. There seems little doubt that the particles in human blood that he saw with 'a very good microscope' were red cells. In the case of Leeuwenhoek, the discovery of red cells can be precisely dated. In letters communicated to the Royal Society[5] on 15 August 1673 and 17 April 1674, and published bearing the date 27 April 1674 in the *Philosophical Transactions*, Leeuwenhoek says: 'I have divers times endeavoured to see and to know, what parts the *Blood* consists of; and at length I have observ'd taking some blood out of my own hand, that it consists of small round *globuls* [sic] driven through a Crystalline humidity of water. Yet, whether all Blood be such, I doubt. And exhibiting my Blood to myself in very small parcels, the globuls yielded very little colour.' It seems very likely that Leeuwenhoek saw red cells rather than fat globules in the shed blood, but he does go on to say that he saw similar 'globuls' in milk, hair from his own head and nail cuttings from his hand. He did not doubt that the nail grew by protrusion of 'globuls', and this was also true of the hair of an elk which was composed of 'conjoined globules'. In fact Leeuwenhoek came to believe that all animal matter was composed of globules which Christiaan Huygens, who knew as much as anyone at that time about the refractive errors generated by lenses, strongly doubted. Writing to Henry Oldenbourg, on 30 January 1675, Huygens asks 'I should very much like to know what credence is given by your people to the observations of our Mr Leeuwenhoek who turns everything into little balls.'* Huygens himself was unable to see such globules and thought that they might be an optical artefact. At a later stage Leeuwenhoek did, in fact, agree that dentine was composed of tubules, but, whatever might be the case for other tissues, there is no doubt that by 1682 Leeuwenhoek was looking at the red cells in blood. The key observation is described in a letter to Robert Hooke, who gave Leeuwenhoek priority in the discovery of red cells. Dated 3 March 1682, this letter contains the following passage: 'Observing the blood [of a ray] it struck me that the parts of the blood that are globules in human beings and mammals, and colour the blood red, are here all of them flat, oval particles, thickish floating in a crystalline water. Where these oval particles lay single they had no colour, but when three or four lay on top of each other they began to show a red colour.' Dobell, in his study of Leeuwenhoek, claims that, in order to account for the latter's remarkable success as a microscopist, he must have used some sort of specialized illumination akin to dark ground. Leeuwenhoek was certainly very secretive about his microscopical methods. In a letter dated 22 January 1675 to Oldenbourg he says: 'but I can demonstrate to myself globules in the blood as sharp and clear as one can distinguish with one's eyes, without any help from glasses, *sandgrains that one might bestrew upon a piece of black taffety silk*' ('de santgens . . . diemen op een swart sijde taff soude moyen werpen'). But the development of colour when the globules overlapped argues against dark-ground illumination in the strict sense, at least in this instance. A less precise form of oblique lighting seems more probable.

Malpighi's observations on blood certainly antedated those of Leeuwenhoek, but they are much more ambiguous. In his *Exercitatio de omento, pinguedine et adiposis ductibus* (Treatise on the omentum, fat and the adipose ducts), published in 1665, Malpighi describes the contents of peritoneal blood vessels in the hedgehog. In the blood he saw what he thought to be fat globules 'possessing an outline of a particular shape, and reddish; they resembled in general a garland of red corals'. There is little doubt that Malpighi was looking at rouleaux of red cells, and this appears to be the first mention of these objects in print; but Malpighi evidently does not distinguish them from globules of fat.

It occurred to no one that the corpuscles that were seen in blood could have had some homology with the elements that made up solid animal tissues. Plant cells being much larger than those of animals, and in many situations endowed with thickened walls, cross-sections of plant parenchyma often presented the fenestrated or cellular appearance that has been described. But sections of solid animal tissues, such as muscle or liver, presented no comparable appearance. It was therefore natural to suppose that they were organized in some fundamentally different way. Francis Glisson, to whom Grew first sent his work on the microscopic structure of plants and whose *Tractatus de rachitide* (Treatise on rickets) of 1650 and *Anatomia hepatis* (Anatomy of the liver) of 1654 were among the first medical monographs in English, devoted the last years of his life to the analysis of solid animal tissues, and particularly the analysis of muscle. In his *Tractatus de natura substantiae energetica*

(Treatise on the irritable nature of tissue), published in London in 1672, Glisson proposed that the fundamental subunits of solid animal tissues were 'irritable' fibres composed of concatenated atoms. Both the supposed fibrous nature of animal tissues and the 'irritability' of the fibres had important consequences. In the late seventeenth and early eighteenth centuries the notion of 'irritable fibres' was substantially extended by the classification given by Giorgio Baglivi (1668–1707), Malpighi's favourite disciple. Together with a great deal of astrological speculation, Baglivi proposed that the fibres were of two kinds, the *fibra matrix* and the *fibra membranacea*, which presumably corresponded to the muscle system on the one hand and non-contractile elements on the other. Baglivi took the view that the whole body was composed of these two functional elements. But by far the most influential figure in animal physiology in the eighteenth century was the Swiss anatomist Albrecht von Haller (1708–77). In the first volume of his *Elementa physiologiae corporis humani*, published in Lausanne in 1757, he put forward the view that the basic elements of which the body is made up are fibres: 'For the fibre is for the physiologist what the straight line is for the geometrician, and from this fibre all shapes surely arise.'* Haller believed that the larger fibres that one observed were composed of smaller ones, which were linear arrays of atoms held together by gluten and moulded by pressure. The nature of the pressures is not specified, but, according to Haller, it was these microfibres that constituted the fundamental subunits of animal tissues. Haller divides the microfibres into three categories. First, there is the *tela cellulosa* (cellular tissue) which forms the framework that supports the body. As pointed out initially by Turner in 1890[6] this refers specifically to connective tissue, and not to the body as a whole. Second, there is the *fibra muscularis* which is endowed with intrinsic 'irritability'; and third, the *fibra nervosa* endowed with sensibility. To these three categories of fibre, Haller attributed the various functions that characterized animal organs; but he says little about the microscopic anatomy of epithelia or of glandular tissues. Numerous authors in the eighteenth century accepted Haller's classification, or at least the assumption that fibres were the ultimate constituents of animal organs. This view reached its apogee in the work of Marie François Xavier Bichat (1771–1802) who might be regarded as the founder, or at least the key figure, in the indigenous French school of animal histology

The 'irritability' that Haller ascribed to the *fibra muscularis* left its legacy in the work of subsequent generations. Fontana (1730–1805), about whom more presently, was greatly interested in the phenomenon of 'irritability', and so was Leopoldo Marco Antonio Caldani (1725–1813); for a brief period the two collaborated. But it was Luigi Galvani (1737–98) who established that 'irritability' had its origin in the conduction of an impulse from nerve to muscle. In his classical experiments on 'animal electricity', described in the *De viribus electricitatis* (On the powers of electricity), published in Modena in 1792, Galvani showed that in an isolated frog nerve-muscle preparation, contraction of the muscle could be elicited by the passage of electricity. These

11. Albrecht von Haller (1708–78)

experiments laid the foundation stone for modern electrophysiology which was given its basic methodological infrastructure by Johann Wilhelm Ritter (1776–1810), Alexander von Humboldt (1769–1859) and, eventually, Emil Du Bois-Reymond (1818–96).

An eighteenth-century view that could be regarded as an alternative to Haller was presented by Kaspar Friedrich Wolff (1733–94), who began life in Berlin and ended it in St Petersburg as professor of anatomy in the Academy of Sciences and director of its botanical garden. In many respects Wolff's model for the formation of animal tissues subsumed that of Haller. The *Theoria generationis*,[7] suffused as it is with notions of a philosophical or naturphilosophical kind, drew a parallel between animal and plant tissues

12. Xavier Bichat (1771–1802)

13. Kaspar Friedrich Wolff (1733–94)

which in some respects antedated the parallelism asserted later by Dutrochet, Raspail and Turpin. The model advocated by Wolff for both animal and plant tissues leaned heavily on the ideas of Grew. Wolff considered that the fundamental subunit of all tissues was a vesicle or globule which, like Hooke and Grew, he sometimes called a cell. Fibres and vessels he regarded as secondary structures. Like Grew he imagined these vesicles to be spaces formed in the growing tissue and initially containing air. Later, some of them were filled with fluid, which Wolff regarded as a mechanism for the storage of excess sap. Similarly, fat cells in the animal body were thought to be depots for the storage of excess nutrient. Wolff noted that in plants there were larger vesicles and smaller ones interposed between them, and concluded from this that the growth of the whole plant was due to the production of new vesicles: 'The leaves therefore grow for the most part by means of new vesicles interposed between the old ones, but partly also by the distension of existing vesicles. And the branches and trunks similarly enlarge, partly by the interposition of new vessels among those that had so far made up these structures and partly by the distension of old vessels.'*

Wolff believed that the production of new vesicles could take place in the absence of fibres and vessels, as illustrated by the seed leaf: 'The young leaf arising from the seed turns out to be composed entirely of vesicles and is patently quite devoid of fibres, vessels or grooves of any kind.'* The new vesicles simply arose *de novo* within a glassy ground-substance: '. . . They arise from a hitherto unadulterated, homogeneous, glassy substance, without any trace of vesicles or vessels.'* The second part of Wolff's book deals specifically with animal tissues, but provides little in the way of new observational

14. Felice Fontana (1720–1805)

evidence. Wolff assumes that animal tissues are constituted in essentially the same way as plants: 'The constituent elements of which all parts of the animal body are at first composed are globules which can always be made out with a modest microscope. If, however, someone were to say that he couldn't see a body on account of its small size, would not its constituent parts escape notice on account of their very smallness?'* Studnička,[8] who made a special study of the resolving power of the microscopes that Wolff might have had at his disposal, comes to the conclusion that Wolff could not actually have seen animal cells.

A third model, which attracted great attention at the time, was proposed by Felice Fontana (1720–1803). Fontana was called to Florence by Leopold I, Grand Duke of Tuscany, in 1776, in order to develop the study of physics, and in the first instance worked in the Pitti Palace. Later he founded and directed a museum of physics and natural science which contained his own wax models. Although an abbot, he supported the French Revolution and was imprisoned in 1799. He lies buried in the church of Santa Croce. The first part of Fontana's major work was published in Italian in 1765, but a more complete version in French appeared in 1781.[9] An English translation of the French edition was made by Joseph Skinner in 1787.[10] This treatise is primarily concerned with snake venom and various vegetable poisons, but, as its lengthy subtitle indicates, it includes a section on the basic structure of the animal body. Fontana believed that all animal tissues, with the possible exception of the membranes of the vitreous and crystalline humours, were composed of twisted cylinders – 'cylindres tortueux primitifs'. He saw these structures in nerves, tendons and muscles, which contained the smallest cylinders that he could resolve, but he was convinced that such structures were the fundamental subunits of animal tissue, even when microscopic observation failed to reveal them. 'The primitive twisted cylinders that I found in the cellular tissue of nerves, tendons and muscles are the smallest to be found in all the parts and organs that I know of. They are much smaller than the smallest of the red vessels which let through only one blood corpuscle at a time. All attempts I have made to break them down into cylinders of smaller size have failed.'* And then the extrapolation: '. . . I know of no part of the animal body that doesn't show twisted cylinders if it contains cellular tissue.'* 'Cellular tissue' again probably means connective tissue in this context, and it is possible that what Fontana actually observed were connective tissue fibrils, but it is surprising that he makes this extrapolation to all other tissues, including epithelia, where his own drawings clearly show globular structures and, in at least one case, structures that to a modern eye are easily recognizable as flattened epithelial cells.

Fontana's views were challenged by Alexander Monro[11] who regarded twisted cylinders as optical artefacts, but the celebrated Jena anatomist, Carl Friedrich Heusinger, in his *System der Histologie*,[12] treats Fontana with respect. Although the idea that animal tissues were composed of twisted cylinders attracted a good deal of attention at the time, it was not widely accepted and appears to have had no modern repercussions. But it is probably true, as will be discussed in more detail in a later section dealing with the discovery of the cell nucleus, that Fontana was the first to see a nucleus in a cell other than a red cell.

CHAPTER 3

The Case for France

It is a sad fact that nineteenth-, and indeed, twentieth-century accounts of the origin of the cell doctrine are beset with nationalistic prejudices that exaggerate the contributions of one's own country and disparage the efforts of others. Nowhere is this better seen than in historical accounts emanating from France and Germany. As might be expected, scientific writing is very much the product of its time, and often accounts of scientific discoveries are prefaced by historical introductions that clearly reflect the politics of the day. For example, in his *History of Botany*, Julius Sachs[1] voices the opinion that, in deciphering the nature and function of cells, only Germans are worth considering ('. . . nur von Deutschen die Rede ist'). The English were said to have contributed little since Grew; the French, though they wrote a lot, were judged not to have solved any fundamental problems. Behind Schwann, there stood a cohort of German scientists who not only gave him priority in establishing the cell doctrine, but accepted, without any experimental evidence, the model he proposed for cell generation. Von Baer marred a classical paper by a completely unjustified disparagement of the work of Rusconi, and virtually all subsequent German writers attribute to von Baer a priority that actually belongs to Rusconi. Schleiden, in his usual aggressive manner, has this to say about Raspail: 'To go into Raspail's work seems to me unworthy of science. Anyone who feels the inclination to do so can look the man up himself.'[2]* And even in 1927, Studnička,[3] writing from Brünn in German, produced an historical account that is both meticulous and exhaustive in arguing the case for Purkyně, as opposed to Schwann, but he does no more than mention Raspail's name as a possible forerunner. In a later article, written in 1931[4] and dealing specifically with precursors of the cell theory, he gives Raspail more space, but, even so, hardly does him justice.

On the other hand, writers in the French language, and especially those steeped in the Gallic tradition, naturally spring to the defence of French science and often exaggerate the importance of contributions made by French scientists. For example, Paul Broca, in the historical introduction to his *Traité des tumeurs*[5] says 'It is generally thought that the cell theory is a German concept; that is completely wrong. It was born not in 1838, or even in 1837; it is not the brain-child of either Schwann or Schleiden. It is twelve years older than that; it is French and is the property of M. Raspail.'* Marc

Klein's monograph *Histoire des origines de la théorie cellulaire*[6] also misses no opportunity to stress the primacy and the importance of French contributions. And Duchesneau, in his book *Genèse de la théorie cellulaire*,[7] which appeared in 1987, begins with sixty-five pages devoted to Dutrochet and Raspail and ends with the conclusion: 'As a consequence of the propositions gathered about the idea of a cell, the theories of Dutrochet and Raspail, ahead of that put forward by Schwann, served as an experimental programme as early as the eighteen-twenties.'*

This national rivalry is eminently understandable given the political tension that developed between France and Germany in the nineteenth century; but it is not helpful in assessing the relative contributions of French and German scientists. There was, of course, the general upheaval in Europe precipitated by the French Revolution, but 1806 brought the humiliating defeats of Jena and Auerstedt which crystallized the German revanchism that culminated in the occupation of Paris in 1814. Throughout the nineteenth century the rise of Prussian nationalism was accompanied, in university circles, by what in Adlerian terms might reasonably be called an inferiority complex. This manifested itself as a form of aggression which sought to prove that German academic prowess was at least the equal of, if not superior to, that of other countries, and especially France. The battle of Königgrätz (Sadowa) in 1866 and, even more, the proclamation of the German Empire at Versailles in 1871, set the seal on German claims to superiority, and scientific papers in the German language during this period do less than justice to the contributions made elsewhere. Citations in the literature are often heavily biased in favour of discoveries made by Germans, and reference to French papers is often needlessly disparaging or omitted altogether. Further examples of this reciprocal nationalistic bias, which are not hard to find, will be given in later pages.

The other difficulty that arises in assessing the contributions of individual authors is whether they were right or wrong, or at least how far they were right or wrong. This is a major preoccupation of scientists and more often than not separates them from professional historians. Many professional historians seem to have adopted Popper's view that we can only be provisionally right, and that we are all wrong, at least to some degree, in the end. This point of view usually leads to a consideration of scientific work against a contextual background of other, often social, factors, and a relative indifference to the correctness of what has been observed. For example, Schiller and Schiller, in their biography of Dutrochet (1776–1847)[8] refer to the detailed studies of Baker as an example of the imposition of modern orthodoxy on past events, and many historians consider the simple distinction between right and wrong to be unhistorical. But it is difficult for a scientist writing about past events to ignore their relevance to reality. We cannot give equal weight to Harvey's arguments concerning the circulation of the blood and those of Riolan who opposed him. Riolan's arguments were very much the product of the intellectual and cultural background in which he found himself, but we lend much more weight to Harvey's views because the blood, after all, does go round and no contemporary scientist believes that the circulation of the blood will ever be disproved. In considering the works of individual authors, therefore, a serious attempt will be made to discuss them in their historical context, but the present author does not believe that the distinction between being right and being wrong is necessarily unhistorical, and makes no apology for drawing this distinction from time to time when the occasion presents itself.

At the end of the eighteenth century the dominant medical figure in France was François Xavier Bichat who served as a physician at the Hôtel-Dieu in Paris and died in 1802 at the age of thirty-one. Bichat is now regarded in French cultural tradition as one of the founding fathers of modern histology: a hospital and, more recently, a university department have been named after him. But in terms of the long and broad view of European history, Bichat's model of the animal body was essentially an extension of the ideas of Haller. For whereas Haller reduced animal organs to three kinds of basic fibre, Bichat, in his *Anatomie générale*[9] of 1801 distinguished twenty-one different fibres. These could be divided into two classes: simple fibres and compound fibres woven together ('tissus') in a combinatorial manner that conferred specificity of function. In Bichat's view, function was determined not only by the anatomical position of an organ and its shape, but by its basic structure, and he envisaged metabolic specialization to be the consequence of the detailed architectural differences in particular tissues. It must, however, be mentioned that Bichat viewed microscopic observations with the greatest reserve: he was only too aware of the optical artefacts that microscopes generated. He therefore refused to use a microscope at all, and his observations were limited to what could be discerned by teasing, dissection, maceration and a hand lens. I think it is true to say that he made no contribution to what we now know as the cell doctrine, and although in the *Anatomie générale* he devotes many pages to the 'système cellulaire', it is clear once again that the reference here is to connective tissue. There is a sense in which the ideas of Bichat form a basis for modern tissue physiology, but they certainly do not form a basis for modern microscopic anatomy.

We have now to consider four names, of which two at least are major figures in their own right and frequently cited by French historians as important precursors of the cell doctrine: Milne-Edwards, Dutrochet, Raspail and Turpin. Henri Milne-Edwards, despite his English name, was born in Bruges and eventually became a professor of zoology at the Muséum national d'histoire naturelle in Paris. As far as the cell doctrine is concerned, there are two communications of his that need to be considered: the thesis deposited on 30 July 1823 at the Faculty of Medicine in Paris,[10] and a later article presented on 19 August 1826 to the Société philomatique and published in the following year.[11] The thesis appeared one year before Dutrochet's book on the fine structure of animal and plant tissues, the *Recherches anatomiques et physiologiques* of 1824,[12] and there is evidence that the two men exchanged ideas that were incorporated in Milne-Edwards's thesis. However, it is clear that in his own work Dutrochet largely adopted the model proposed by Milne-Edwards, even though he did modify it in many important respects.

The 1826 article by Milne-Edwards is of less interest than his thesis of 1823 and adds little to it. It is difficult to be sure just what it was that Milne-Edwards saw. He used an achromatic compound microscope, i.e. one largely free from colour artefacts, made by Adams, which was one of the best available at that time. Studnička[13] was convinced that he could not have seen the cells of animal tissues, not because the instrument was incapable of it, but because appropriate methods for the preparation and examination of animal tissues had not yet been devised.

Milne-Edwards did, however, see red blood cells and even the central shadow within them, but he did not equate these cells with what he assumed to be fundamental subunits that he saw elsewhere. What he claims to have seen in other tissues was an array of globules uniformly 1/300 mm in diameter. It is difficult to understand why Milne-Edwards insisted on the uniform size of his globules. If he was simply seeing halations induced by the illumination that he used, one would have expected that these halations would vary in size, and that he could hardly have avoided noticing this. It was perhaps the conviction, inspired by some sort of atomism, that the basic components of animal tissues had to be uniform in size that induced Milne-Edwards to see what he wanted to see.

He began by examining the subcutaneous 'tissu cellulaire' (connective tissue) of the human thorax, presumably in the unfixed and unsectioned state. Like Fontana before him, he saw 'des cylindres tortueux'. These presumably corresponded to bundles of collagenous fibres, but Milne-Edwards claimed that each of these fibres was composed of an irregular but linear concatenation of uniform globules. This he confirmed in the connective tissue of a dog's paw, in bovine peri-aortic tissue, in the subcutaneous connective tissue of a cockerel and a frog, and in connective tissue taken from the abdominal cavity of a carp. 'Suffice it to say' he writes 'that the fine structure seemed to me to be everywhere the same, and the elementary globules, in their shape and in their diameter, similar to those that are to be seen floating in pus, in milk and so on.'* He found similar arrays of uniform globules in mucous and serous membranes and also in voluntary muscle, where Swammerdam and a number of other earlier writers had already described myofibrils and the presence of cross-striations within them.

Milne-Edwards first examined human femoral muscle and then muscle from a range of mammalian, amphibian, piscine and even invertebrate tissues. He also examined skin and nervous tissue including both grey and white substance of the brain. Everywhere he found his uniform globules and concluded that in all animals the basic structure of all tissues is 'a rather large number of these corpuscles which might differ in their constitution, but which vary but little in their shape and probably in their volume'.* The

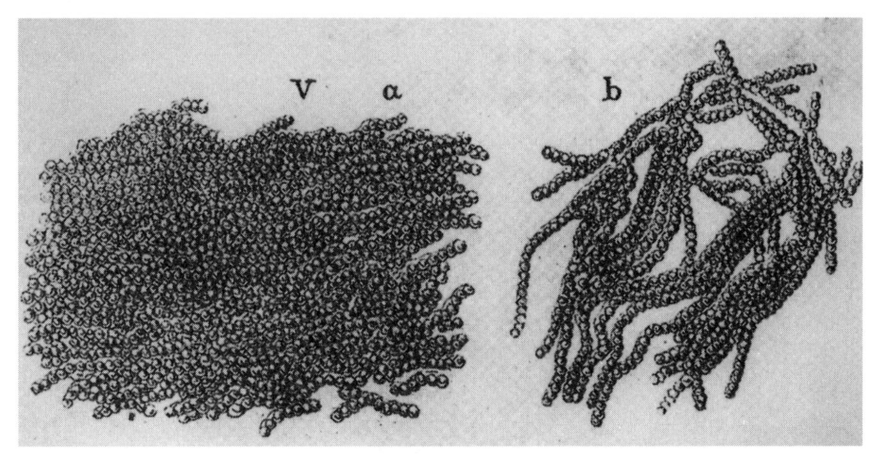

15. Arnold's view of the fine structure of epidermal tissue (a) and muscle (b). Both are composed of arrays of uniform globules that differ only in their disposition

problem posed by the now well-known and well-defined fine structure of plants, in which the constituent cells obviously differed in size, Milne-Edwards overcomes by proposing, as Wolff had done before him, the generation of large structures from much smaller globules that might be beyond the reach of the microscope: 'These globules, which might be called elementary, are perhaps themselves formed by still smaller corpuscles that our present methods of investigation do not permit us to see.'*

Milne-Edwards's views were not only widely known and quoted in France; he also had the support, at least provisionally, of many other investigators across Europe. In Arnold's *Lehrbuch der Physiologie des Menschen*,[14] one of the most widely read German texts at that time, Part 1 contains diagrams of the fine structure of muscle and epidermal tissue, both composed of linear arrays of uniform globules that differ only in their disposition, being more closely apposed in one case than in the other. H. E. Weber, however, whose *Allgemeine Anatomie des menschlichen Körpers*[15] went through four editions by the year 1830, was extremely critical of the observations made with contemporary microscopes. His book contains an extensive review of microscopal techniques, and he is fully aware of the false images produced by errors in refraction. He regarded Milne-Edwards's observations as an example of optical artefact. To a modern eye, Milne-Edwards was simply mistaken, but to his contemporaries the model that he advanced was immensely attractive because it appeared to provide decisive evidence for a basic, uniform, 'atomic' structure that almost all biologists at that time were convinced must exist.

Henri du Trochet, which he turned into Dutrochet after the French Revolution in order to draw attention away from an aristocratic background that went back to the fourteenth century, is in a different category altogether.

16. Henri Dutrochet (1776–1847)

Most French accounts of his work are marred, in my view, by an excessive preoccupation with his relationship to Schwann. Schwann does not cite him, although he must have been aware of his contributions, and this has provoked several authors, notably A. R. Rich,[16] to argue that the main credit for the formulation of the cell doctrine belongs to Dutrochet and not to Schwann. This is a question that cannot be avoided, and it will be considered in some detail at a later stage when Schwann's *Mikroskopische Untersuchungen*[17] is discussed. However, Dutrochet is a figure of the first importance, and his discoveries should be examined quite independently of those subsequently made by Schwann. Schiller and Schiller[18] consider that Dutrochet's work was dominated by two fundamental ideas: a materialistic view of the phenomena of life, to be seen in his scientific philosophy as well as in his experimental research; and the identity of vital phenomena in animals as well as plants. To quote from the *Mémoires* of 1837[19] '*Life is one*; the differences shown by its various phenomena, in all things that are alive, are not fundamental differences; if these phenomena are tracked down to their origins, the differences are seen to disappear and an admirable uniformity of plan is revealed.'* Of course, one cannot compare the sophistication of Dutrochet's views on the fundamental uniformity of life-forms with those propounded almost a century earlier by Wolff, but both authors did think that animal and plant tissues were composed essentially of vesicles or globules, and both concluded that this was so despite the fact that, with one possible exception in the case of Dutrochet, neither author had actually seen animal cells as we now know them.

Now Dutrochet was not alone in having a materialistic view of life, and he was certainly not the first to propose that animal and plant tissues were constituted in the same way. Nor was he the first to disaggregate plant tissues into their component cells. Moldenhawer in his *Beiträge zur Anatomie der*

Pflanzen,[20] which, although published in 1812, embodied some eighteen years of work, had already described a procedure for isolating plant cells by macerating the plant in water. Moldenhawer's technique was in fact more gentle, and hence more widely applicable, than the hot nitric acid procedure devised by Dutrochet. However, Dutrochet was too good an observer to fail to see that the cells he isolated from plants differred in size, and he was therefore inclined to dissent from Milne-Edwards's contention that the fundamental subunits of animal tissues were uniform in size. But, on the whole, Dutrochet accepted Milne-Edwards's model, and the outstanding insights that Dutrochet provided into the physiology of plant and animal tissues rested on a view of these tissues as aggregations of membrane-bound globules or vesicles. Nowhere is there any mention of a cell nucleus.

In my view, Dutrochet's originality lay neither in his philosophical position nor in his insistence on the similarity in the cellular structure of animal and plant tissues. What no one appears to have done before him is to consider cells as physiological entities, even though his anatomical vision of them was, by present standards, defective. He considered cells as the basic units of metabolic exchange, their nutrition mediated by the selective inflow of compounds across the cell membrane and their waste products eliminated by the selective outflow of other compounds. He developed the concepts of endosmosis and exosmosis (the passage of molecules into and out of the cell against the concentration gradient) and coined these terms to describe the phenomena that he had discovered. He provided strong evidence for the view that endosmosis was not a matter of capillarity or viscosity, nor did it appear to be mediated by trans-membrane electric currents. These experiments were done with organized tissues, but the results were always extrapolated to the cell, and the semi-permeability of the cell membrane was always envisaged as the primary determinant of the molecular flow: 'This uniformity of fine structure proves that organs actually differ only in the nature of the substances contained by the vesicular cells of which they are entirely composed; it is in the cells that the fluid appropriate to each organ is secreted.'* And again: 'If one compares the extreme simplicity of this astonishing structure [the cell] with the extreme diversity of its innermost nature, it is clear that it constitutes the basic unit of the organized state; indeed, everything is ultimately derived from the cell.'*

The other leitmotiv of Dutrochet's work was the study of movement in plants as well as animals. Haller had considered that the property of 'irritability' was confined to animals, but Dutrochet, impressed above all by the behaviour of the sensitive plant (*Mimosa pudica*) regarded movement as a characteristic of living forms, both plant and animal. He studied the movement of roots, stalks and leaves in the sensitive plant, the translocation of sap, intracellular flow in *Chara* and muscular contraction in the animal. All these he sought to reduce to the one general principle: a change in tissue curvature produced by the endosmotic inflow of fluid. This view gave rise to controversy in the Académie des sciences, but it initiated a materialistic and experimental approach to phenomena that were largely shrouded in speculation and not a little mysticism. Endosmosis and the study of movement formed the essential pillars on which the edifice of general physiology was eventually erected. It could reasonably be argued that this contemporary science, and its consistent attempts to explain physiological phenomena in cellular terms, owes its origin to Dutrochet as much as to anyone else.

Although Dutrochet must be regarded primarily as a cell physiologist, his contributions to microanatomy were not negligible. Like many others at that time, he was sceptical about the optical artefacts produced by compound microscopes and in the *Recherches anatomiques* of 1824[21] recommended the use of a simple microscope or a loupe. This was three years before Amici introduced achromatic lenses in France, by which time Dutrochet had given up anatomical investigations altogether and turned to cell physiology. In considering Dutrochet's anatomical work we are therefore confined to his early investigations and not those which formed the centrepiece of his mature work. He was sixty-one years old when Schwann deposited the first two sections of the *Mikroskopische Untersuchungen* with the Académie des sciences in Paris, and he was by then only moderately interested in questions of fine structure. In his *Mémoires* of 1837,[22] only the last chapter is concerned with the elementary compositon of animal tissues, and there is no mention of the substantial contributions made by scientists in Germany during the 1830s.

There is, however, one observation, made with a simple microscope, that may confer on Dutrochet an important priority even in the field of microanatomy. In examining the brains of the gastropods *Helix pomatia* and *Limax rufus*, he noticed and illustrated large globular structures that were surrounded by many smaller, ovoid or spherical, corpuscles. Studnička,[23] who studied this very material with a microscope attributed to Dutrochet, came to the conclusion that what Dutrochet saw were ganglion cells surrounded by microglia (small accessory cells in the nervous system). The ganglion cells in the brains of these molluscs are known to be very large and can apparently be observed with a simple lens. In any case, if Dutrochet actually saw what he illustrated, and there is no substantial reason to doubt it, he was probably the first to observe a genuine animal cell, even if an exceptionally large one. For other animal tissues, which he found too opaque for microscopical examination, he adopted, as has previously been described, a modified version of Milne-Edwards's model which, however, did not lead him seriously astray in the matter of cell physiology.

Dutrochet believed that tissue growth was achieved by the formation of new globules, and, being a convinced opponent of spontaneous generation, he was not attracted to the idea that they arose *de novo* in the fluid that surrounded them. He thought, on the contrary, that new globules were formed within old ones, a process that later authors called endogenous cell generation. But since he had no notion of the existence, to say nothing of the function, of the cell nucleus, Dutrochet's speculations on the mode of formation

17. François Raspail (1794–1878)

of new globules did little to advance knowledge in this area. Indeed, it appears that at one stage he took what were obviously plastids associated with the plant cell wall to be newly formed globules. The main difference between the model proposed by Dutrochet and that proposed by Schwann was that Schwann, under the influence of Schleiden, put the cell nucleus at the centre of his scheme for cell generation, although this scheme turned out to be no less misleading than that of Dutrochet. Dutrochet's work was well known in Germany. Studnička records that his own copy of the *Recherches anatomiques* bears an inscription in Dutrochet's hand dedicating the book to Heusinger.[24] And, at a later date, Du Bois-Reymond[25] couldn't speak highly enough of Dutrochet's work on endosmosis. It remains surprising that Schwann does not quote him.

Perhaps more than any other French writer in the 1820s and 1830s, François-Vincent Raspail (1794–1878) is given short shrift by German microscopists, although none is as disparaging as Schleiden. There seem to be three specific reasons for this, quite apart from the prevailing nationalism. The first was that for most of his life Raspail's scientific work was virtually unknown or ignored, even in France. As an intransigent republican he was imprisoned under the July monarchy and again after the turbulence of 1848. Indeed his *Nouveau système de chimie organique* of 1823[26] was revised in prison. He spent a decade, from 1853 to 1863, in exile in Brussels, and on his return to France, although elected a deputy in 1868, was imprisoned again. He was re-elected as a republican deputy in 1876 but it was not actually till after his death in 1878 that his reputation was resurrected. Reputation in due course became fame, and there is now, among other memorials, an elegant Boulevard Raspail in Paris.

The second reason for Raspail's scientific work being largely ignored is that in mid-career he became interested in clinical problems and devised therapeutic regimes that had great popularity, and survive even to the present day, but had little, if any, efficacy. In this, Raspail to some extent resembled Robert Remak, who will be considered later, although the similarity is in other respects strained. And the third reason is that German microscopists failed on the whole to appreciate the genuine and highly original contributions that Raspail did make to our understanding of the cell. German anatomists were primarily concerned with establishing, by microscopic means, the precise structure of the cell and its mode of formation. Raspail, on the other hand, was primarily concerned with the cell's chemistry. Like Dutrochet, he made certain assumptions about cell structure and cell formation, but it did not greatly affect his experimental work that some of these assumptions were wrong.

Raspail's microscope, made by Deleuil of Paris, contained a biconvex lens which gave a magnification of ×20; a tourmaline lens which gave a magnification of ×240 and reduced chromatic aberration by polarizing the light; and a double lens that was composed of two blue glass plano-convex lenses and gave a magnification of ×50. This was an excellent instrument for its time and should have delivered exact images. Nonetheless Raspail envisaged cells much as Dutrochet had done, although he did not accept many aspects of Milne-Edwards's model. He did not, for example, believe that the cells, or globules, were uniform in size or that they were joined together in linear arrays. He found them scattered among fibres and membranes, and he did not think the latter were corpuscular in nature. He saw cells in the epidermis and dermis, and red cells in the blood, although he thought the cells in the blood differed in size from those in other organs. There is no doubt that he delineated nerve cylinders, and he concluded that, like muscle, they were composed of aggregated fibres. He speculated that nerve cylinders were produced by elongation of cells and muscle fibrils by gigantic cellular enlargement.

Although some of his anatomical observations proved eventually to be inaccurate, Raspail was overwhelmingly convinced of the central role of the cell in organized living forms. This conviction is encapsulated in a paraphrase of Laplace that has now become immortal: 'Give me an organic vesicle endowed with life and I will give you back the whole of the organized world.'[27]* To quote from a passage reproduced in Dora Weiner's biography of Raspail:[28]

> Each cell selects from the surrounding milieu, taking only what it needs. Cells have varied means of choice, resulting in different proportions of water, carbon and bases which enter into the composition of their walls. It is easy to imagine that certain walls permit the passage of certain molecules, others condense on their external walls. You can see how varied the result can be, . . . according to the number of carbon and oxygen molecules – six, eight, twelve – surrounding the central carbon molecules. The modifications are infinite. A cell is therefore a kind of laboratory within which all tissues organize and grow.[29]

Note, however, that there is still no inkling of a cell nucleus.

On the genesis of cells Raspail's views did not differ greatly from those of Dutrochet. He believed that new vesicles were formed within old ones. This idea initially led him into an error that greatly exercised later botanists. In one of his earliest works, *Développement de la fécule* (1825)[30] he argued that starch grains were the progenitors of new vesicles, and he reiterated this view in his *Nouveau système* of 1833.[31] The starch grains were said to enlarge until they touched each other; they then burst and released the vesicles that they contained. These vesicles were supposed to contain still smaller ones, a process that Raspail called 'emboîtement' and that we now associate with nested Russian dolls. The newly formed vesicles were thought eventually to occupy all available space, and in this way Raspail derived the whole of the leaf, with the exception of the midrib and the veins. Stems were said to be similarly formed; and, without much evidence, the model was extended to animals. At one stage Raspail believed that the cell wall contained both male and female globules, and that the cell was thus essentially a hermaphrodite; he looked for specific generative organs within the globule itself. The arguments that later botanists advanced in support of, and against, the generative role of starch grains and other plastids will be discussed in Chapter 4.

In the *Développement de la fécule* there is a Latin aphorism[32] that is of especial historical interest. As an epigraph, Raspail uses the phrase 'Omnis cellula e cellula.' (Every cell is derived from another cell.) As far as I am aware, this is the first use of this particular phrase, although, as will be shown later, variants, perhaps beginning with Harvey and passing through von Siebold and Oken, crop up repeatedly in the scientific literature of the nineteenth century. The phrase is usually associated with Virchow, although Virchow and Raspail had quite different views about the mechanism of cell formation. Virchow was far from generous in his references to his predecessors, and the question arises whether he knew about Raspail's prior use of the phrase when he adopted it as the leitmotiv of his *Cellularpathologie*. Oken, who founded the Gesellschaft deutscher Naturforscher und Ärzte (Society of German Naturalists and Doctors) wrote to Raspail in 1832 asking him to summarize, for publication as an abstract in *Isis*, the work that had appeared in the *Mémoires de la société d'histoire naturelle de Paris*. Oken's letter is very flattering but perhaps indicates that he was not too familiar with Raspail's work. Whether Raspail knew about Oken's work, and in particular *Die Zeugung*[33] which appeared in 1805, is an open question, for *Die Zeugung* concludes with two aphorisms: 'Nullum vivum ex ovo!' (No living thing comes from the egg), in contradiction to what Oken believed to be the dictum used by Harvey; 'Omne vivum e vivo!' (All living things come from other living things). There is one piece of evidence that makes it unlikely that Virchow actually plagiarized Raspail. If he had done that, it is reasonable to suppose that he would have written 'Omnis cellula e cellula'; but the version of the phrase first used by Virchow was 'Omnis cellula a cellula' (the ambiguous ablative meaning 'separate from' or 'by means of') and it was not until Leydig in his *Textbook of Histology* of 1857[34] amended 'a cellula' to 'e cellula' (unambiguously meaning 'derived from another cell') that the adage which has become so familiar was finally adopted.

So much for Raspail's contributions to microscopic anatomy. Although, when they noticed Raspail at all, it was these that drew the attention of German microscopists, anatomy did not form the main theme of Raspail's scientific output. It was his systematic investigation of the chemistry of the cell that proved decisive; and it is no exaggeration to claim that the modern science of cytochemistry originated with him. He was certainly one of the earliest to examine the response of the cell contents to a range of chemical reagents and to do this under the microscope; and he devised experimental methods, some of which are still in use, that made this possible. He was probably the first to freeze tissues in order to make frozen sections, and he was certainly the first to do so with a view to obtaining chemical information about the cell. He invented a form of micro-incineration, burning cells on a platinum spoon and then subjecting the residue to chemical analysis, and he developed a novel range of chemicals to identify individual cellular components. In a very early paper he demonstrated the use of iodine to stain starch grains blue; and, over a number of years, he found reagents that identified albumin, sugar, silica, mucin, resins, calcium oxalate, carbonates, chlorides and iron. One has only to look at the full titles of his major works to see that it was cell chemistry that was central to his research interests: *Recherches chimiques et physiologiques* [note that *chimiques* precedes *physiologiques*] *destinées à expliquer non seulement la structure et le développement de la feuille, du tronc, ainsi que les organes qui n'en sont qu'une transformation, mais encore la structure et le développement des tissues animaux*[35] (Chemical and physiological experiments designed to explain not only the structure and development of the leaf, the trunk, and the organs that are merely transformations of them, but also the structure and development of animal tissues); or *Nouveau système de chimie organique, fondé sur des methodes nouvelles d'observation*[36] (New system of organic chemistry based on new methods of observation); or *Nouveau système de physiologie végétale et de botanique fondé sur les méthodes d'observation, qui ont été dévelopées dans le nouveau système de chimie organique, accompagné d'un atlas de 60 planches d'analyse dessinées d'après nature et gravées en taille douce*[37] (New system of plant physiology and botany based on methods of observation developed in the new system of organic chemistry, with an atlas of 60 analytical plates drawn from nature and reproduced in copper-plate).

I have quoted above a passage in which Raspail calls the cell a kind of laboratory in the sense that it is constantly engaged in the work of balancing catabolism and anabolism; but the cell was also a micro-laboratory for Raspail himself, and much of his productive scientific life was devoted to the design of micromethods that made such a micro-laboratory possible. This work is not belittled by Raspail's mistaken assumptions about the microscopic anatomy of plants and animals.

Turpin is a lesser figure. This, I think, is because his contributions consisted more in an inventive proposal for a change in terminology than in the discovery of new experimental facts. He is mentioned by Schwann but, as appears to be almost the rule for French writers, he is treated with contempt by Schleiden. Turpin accepted, in general, the view that both animal and plant tissues were composed of agglomerations of vesicles, but he did not envisage them as empty or filled with air. He proposed that they were collections of a substance which he named 'globuline' and which he regarded as the basic material of life. Unfortunately he used the word 'globuline' rather vaguely to describe both the vesicles themselves and their contents. Thus, he called unicellular plants 'globulines' and noted the presence of both white and green globulines. In multicellular plants, the green globulines could enlarge into what were commonly termed 'cells' and would then lose their colour. He stressed the essential similarity between unicellular and multicellular forms, but that bridge had already been crossed by Oken in 1805.[38] It has been argued by Klein[39] that Turpin envisaged the vesicles operating as independent metabolic units even after they had been aggregated into multicellular forms, whereas Oken is alleged to have thought that their individuality was submerged in a collective form of organization when they were brought together. I think this distinction is less radical than it may sound. It is likely that both men regarded the vesicles as centres of metabolic activity, and the difference in emphasis seems to be largely a matter of the degree to which this activity was coordinated.

Turpin classified the vesicles composed of globuline into three types: (1) 'Globuline vésiculaire solitaire', which embraces unicellular organisms; (2) 'Globuline vésiculaire enchainée', which refers to globulines aggregated together. These two globulines 'se développent à vu dans la nature' (you can see them everywhere); and (3) 'Globuline captive',

> because it is enclosed within mother-vesicles, which themselves began as Globuline. *Globuline captive* has the same characteristics as the first two kinds of Globuline: the same range of shapes and colours and the same mode of reproduction. But it is distinguished by the fact that instead of living and growing separately, it remains within the interior of the mother vesicles. In this situation, its development is hindered so that it loses its globulinous shape, becomes more or less hexagonal, knits together or is grafted together by means of its surfaces, and thus forms a new body of cellular tissue.[40*]

It is clear from this that the form of cell multiplication envisaged by Turpin was the production of cells within cells, a view espoused by Dutrochet and hence no longer original. Turpin published his first account[41] of this process two years after Dutrochet's book *Recherches anatomiques et physiologiques*[42] had appeared. Turpin expounded this principle yet again in 1828,[43] insisting that cell multiplication was driven by a simple and unique mechanism. In the section of the paper entitled 'Des divers modes de propagation végétale' he writes: 'Every propagative plant body has its origin in a selected vesicle that acts as its mother and as its conceptacle, whether this vesicle is a solitary Globuline or belongs to a Bichatia[†] or a silkweed or a pollen grain or, indeed, to the aggregate of Globulines that form the cellular tissue of higher plants.'* At no stage did Turpin envisage cell division and this despite the fact that he wrote another paper in 1828[44] in which he reported that he was struck by the fact that in a variety of plants, the number of vesicles found during the early stages of development was always a multiple of two.

In the last paper of this series, written in 1829[45] Turpin summarizes his views and answers his critics. His resumé begins with the sentence: 'Dès que la matière s'organise elle se globulise' (As soon as matter is organised it becomes a Globuline); and it reaches its climax in the following passage: 'A particle of Globuline, produced by extension of the internal walls of one of the generative vesicles of cellular tissue, is the origin or the germinal seed either of future vesicles in new cellular tissue or of any body that is capable of propagating the species.'* Turpin obviously favoured a generalization of the broadest possible sweep, and it is of interest in tracing the intellectual currents which might have influenced his thinking that his 1827 paper bears an epigraph from Leibniz, the advocate of monads: 'La variété dans l'unité'. But Turpin's variations on the theme of 'Globuline' were very largely exercises in imagination and, unfortunately, they turned out to bear little relation to reality.

† *Bichatia vesiculinosa* was a name invented by Turpin.

CHAPTER 4

Passionate Disputes among Botanists

By the end of the eighteenth century virtually all botanists accepted that plant tissues were largely composed of cells, as Grew and Malpighi had shown. But there were still four questions that remained undecided and a source of controversy: (1) How were new cells formed? (2) Did cells communicate with each other, and if so, how? (3) What did cells contain? (4) Were all tissues, including vessels, made up of cells that had been modified? Once again we see a phalanx of German scientists aligned against a theory that emanated from a Frenchman. They disagreed with each other, but they were unanimous in their opposition to the scheme that was being advocated in Paris. The Frenchman was C. F. Brisseau-Mirbel (later de Mirbel) and his opponents were principally D. H. F. Link, K. Sprengel and L. C. Treviranus. Charles-François Brisseau-Mirbel (1776–1854) first put forward his views on cell formation in 1802.[1] In 1808 he published his *Exposition et défense de ma théorie de l'organisation végétale*[2] and in 1809 he produced a second edition entitled *Exposition de ma théorie de l'organisation végétale*.[3] In 1808 he wrote: 'The first idea, the basic idea, is that the whole structure of the plant is formed by one and the same membraneous tissue modified in different ways. This fact underlies all others.'* 'I begin with the principle that the whole body of the plant is a cellular tissue whose compartments differ in their shapes and dimensions. This simple idea is the basis of my theory.'* As for vessels, Mirbel simply regarded them as elongated cells: 'Les tubes et vaisseaux des plantes ne sont que des cellules très-alongées.' (The tubes and vessels of plants are merely grossly elongated cells.) Only the large spiral tubes that he calls 'tracheae' are an exception to this rule: 'I see only one exception to this law: the case of the tracheae, those narrow laminae twisted into a helix like a corkscrew.'*

On the question of cell formation Mirbel adopts a position that is diametrically opposed to that of L. C. Treviranus. L. C. Treviranus, of whom more later, was convinced that new cells were generated from granules (Körner) which Mirbel called 'sphérioles'. Mirbel, however, proposed that new cells were produced in three precise ways: 'développement super-utriculaire' in which the new cells were formed at the surface of old ones; 'développement inter-utriculaire', in which they were formed between the adjacent walls of old cells, thus pushing the latter apart; and 'développement intra-utriculaire', in which they were formed within the old cells. In the last case the new cells

18. Charles-François Brisseau-Mirbel (1776–1854)

might either themselves form a continuous cellular tissue, and the old cell in which they arose (l'utricule-mère) be resorbed; or the new cells would not form a continuous tissue but remain attached to the wall of the mother cell. While the details of this scheme are simply conjectures or, at best, a reflection of static appearances seen under the microscope, the essential point is that Mirbel is arguing that cells arise from other cells and not from subcellular particles, such as Treviranus's 'Körner', whether they are within the cell or outside it. On the question of what the cells contained, Mirbel's position did not differ all that much from Grew's. Grew believed, as has been described, that the 'cells' were formed by a process akin to the formation of cavities or bubbles in leavened bread. In Mirbel's monograph of 1808 we read: '. . . and finally I realized that this unification, which seems to me to be explicable only by the existence of lateral fibres, arises from the fact that the whole body of the plant is simply a membranous tissue (Grew's bone-lace) which forms empty spaces that vary in shape and in their dimensions. This results in some of them being small regular or irregular cells whereas others become more or less elongated tubes.'* Although Mirbel is somewhat vague on this point, I take his word 'vacuosités' to imply that the cavities or cells were originally full of air and were only subsequently occupied by sap and other substances.

The main bone of contention between Mirbel and his German opponents was, however, whether and how the cells communicated with each other. Mirbel believed that the cells shared a common wall and that this was perforated by countless pores which permitted the free passage of sap from one cell to another. But G. R. Treviranus, as early as 1805,[4] had already provided evidence that this was not so. He found that a thin section of *Ranunculus ficaria*, dispersed under water by a needle, breaks up into small bladders or vesicles (Bläschen) that have no connection with each other; and he argues that all living forms are derived from aggregates of such bladders. 'The origin of the organisation of living material is an aggregate of vesicles, which have no connection with each other. From these vesicles all living bodies are

19. Gottfried Reinhold Treviranus (1776–1837)

formed, and all that they contain eventually undergoes dissolution.'* And again: 'The first thing to be formed in the seed or the egg is a double external covering of which the harder outer layer has retained the name Chorion and the softer inner layer the name Amnion. These membranes are apparent even when the interior of the seed or the egg is still a fluid substance without any visible organisation.'* Baker credits G. R. Treviranus with being the first to disaggregate the plant into separate cells, but most German writers attribute this discovery to Moldenhawer. It is the case that G. R. Treviranus published in 1805 and 1806[5] whereas Moldenhawer's work did not appear until 1812.[6] However, Moldenhawer states that his book is the culmination of eighteen years' work, and there is no doubt that his methodology was much more systematic and provided a general method that was widely adopted. In Volume 3 of his book, G. R. Treviranus refers to the growth of plants, including the silk-weeds, and he refers to the observations of Vaucher on this material. But he does not envisage that the cells of which the plants are composed (the Bläschen: vesicles) undergo multiplication by binary fission (Theilung). That this idea fails to occur to him despite his familiarity with Vaucher's work lends support to the conclusion that Vaucher did not actually see cell division (he certainly did not describe it), despite the assertion of Hughes[7] to the contrary.

Link (1797–1851), Rudolfi (1771–1832) and L. C. Treviranus (1779–1864) submitted their monographs as prize essays on the subject of plant vessels, in a competition organized by the Königliche Societät der Wissenschaft (Royal Society of Science) of Göttingen. All three essays were accepted, with Link being judged the winner, Treviranus the runner-up and Rudolphi third. Treviranus's essay was, in Sachs's view,[8] undoubtedly the best, and this is recognized by Mirbel, who dedicated his *Exposition et défence de ma théorie . . .* of 1808[9] to M. le Docteur Treviranus. Treviranus's essay *Vom inwendingen Bau der Gewächse.*[10] (On the Inner Structure of Plants) was published in 1806; Link's *Grundlehren der Anatomie und Physiologie der Pflanzen*[11] and Rudolphi's *Anatomie der Pflanzen*[12] in 1807. Karl Asmus Rudolphi, although born in Stockholm, was professor of anatomy and physiology in Berlin and formed part, although perhaps the weakest part, of the German opposition to Mirbel. Rudolphi was the patron and father-in-law of Purkyně, whose major role in the development of the cell doctrine will be discussed later. Link refers to the prior work of Bernhardi,[13] which appeared in 1805, and Sachs regards Bernhardi's monograph as more informative than any of the three Göttingen prize essays. Bernhardi categorically denied that the cell walls were perforated by pores, as proposed by Mirbel, and, in this matter, Link declares himself on the side of Bernhardi and L. C. Treviranus who shared the view of G. R. Treviranus that plant cells were independent, non-communicating units. (G. R. and L. C. Treviranus sometimes published together.) L. C. Treviranus is circumspect in the presentation of his position, but argues it forcefully. As mentioned earlier, he advocates the view that cells develop out of granules (Körner) which are initially found within the cells although he does not rule out the possibility that new cells might be formed from extracellular 'Körner':

> My view that the vesicles collectively making up cellular tissue have their origin in granules that are to be found within cells, is, according to Mirbel, a figment of the imagination. Link's judgement is fairer in that he entertains doubts about my idea and adduces reasons for his doubts. Link's reasons are not decisive, but then I am not inclined to ascribe to my own views the compelling power of truth: rather they are and remain only a probable conjecture.*

The essential piece of evidence that Treviranus gives in support of his model is the group of observations that he has made on the seed-leaves of beans and peas. He points out that the seed-leaves, which are composed of cellular tissue and which are indispensable for the survival of the plant, contain numerous intracellular granules of various sizes. But once the established plant is, say, three-quarters of a foot tall, its cells, except for occasional ones apparently distributed at random, are entirely devoid of granules. '. . . we thus have here before our very eyes a proof that can hardly be disputed, that the granular substance produces the material from which the cells and the vessels of the young plant are formed.'* In further support of this idea, L. C. Treviranus, like G. R. Treviranus before him, refers to observations made on the silk-weed, *Conferva mutabilis*: 'If one examines the growing tips of the branches or the stem of this plant, one again sees that they are replete with granules, but when the branches and stem of the mature plant are examined,

20. Bonaventura Corti (1729–1813)

one finds that they are composed entirely of cells, and within these, new granules are formed.' L. C. Treviranus also refers to the work of Vaucher, but again fails to see in it any suggestion of cell multiplication by binary fission.

Mirbel had argued that the intercellular spaces (lacunae) which one sees in plants were artefacts produced by tearing the cellular material during the course of experimental manipulation. Treviranus comments that no one in Germany doubts the existence of these spaces and suggests that Mirbel, when he has thought more deeply about the question, might come to regret his careless assertion that the spaces do not exist. In his monograph of 1811,[14] Treviranus has come to regard Link's view, that cells are formed extracellularly, as more probable than the idea that they arise within cells, but he does not regard the matter as of overriding importance: 'It doesn't make much difference whether during the transition (from granules to cells) the former maintain their usual granular appearance or whether they are dissolved in a homogeneous fluid, which seems more probable.'* Mirbel, answering aggression with aggression, was no doubt needlessly dismissive in asserting that Treviranus's model was a figment of the imagination, but that, in the end, is what it turned out to be.

But there are two observations that Treviranus made that have stood the test of time. The first was the demonstration that the epidermis was not a membrane, but consisted of a layer of cells; and the second was the rediscovery, made apparently without knowledge of Bonaventura Corti's prior observation, of protoplasmic streaming (cyclosis). Treviranus apparently first saw cyclosis in 1803[15] in *Hydrodictyon utriculatum* and *Nitella* (which he calls *Chara*) *flexilis*. He chose this strain of *Chara* because the common *Chara vulgaris* has a rather thick envelope. The phenomenon was described in print in 1811[16] and again in a collection of papers by G. R. and L. C. Treviranus that appeared in 1817.[17] In an article published in the *Annales des sciences naturelles* in 1827,[18] L. C. Treviranus states that in 1817 he was still unaware of the work of Corti. Corti, who eventually became the rector of the Collegio San Carlo in Modena, gave an excellent description of cyclosis in 1774 in a paper entitled *Osservazioni microscopiche sulla Tremella e sulla circolazione di fluido in una pianta aquajuola* (Microscopic observations on *Tremella* and on the circulation of fluid in an aqueous plant).[19] His observations were first made on a plant that he calls 'Cara' *(Chara translucens minor)* but at a later date he described the phenomenon in plants other than those that grow in water. It is, however, probable that fluid movement in cells was first seen by Fontana, to whose work reference has already been made (p. 21).

D. H. F. Link is even more disparaging in his remarks about Mirbel than Treviranus: 'Brisseau Mirbel, a newcomer, claims to have seen distinct openings, which show up as round holes, in the walls of the cells. No other observer can confirm this.'* Link published his major work in 1807, the *Grundlehren der Anatomie und Physiologie der Pflanzen*,[20] which was followed by a supplement in 1809 and others in 1812. On the question of perforations in cell walls, Link was undoubtedly right and Mirbel wrong. It was, of course, known by 1807 that single cells could be isolated from plants. This had been shown, as mentioned earlier, by G. R. Treviranus and earlier still by Vaucher in silkweeds. This implied that there was a substantial measure of independence in each cell and, further, that adjoining cells were separated from each other by two cell walls and not one, as Mirbel had suggested.

Link adduced further evidence. One often finds, he says, in plants with red spots or stripes, isolated cells filled with red fluid contents surrounded on all sides by colourless cells. Further, if twigs are cut from plants and immersed in coloured fluid, one never sees the movement of this coloured fluid from one cell to another. Link claimed that he had long held the view that where cells abutted on each other they were separated by two partition walls and, indeed, by an intercellular space. This was shown most clearly in a cross-section of the pith of *Datura tatula* where, in the corners, a dark mass could be seen that clearly displayed the intercellular spaces. Link conceded, however, that he had not actually seen a double membrane. In the supplement of 1812 he described further experiments that demonstrated the independence of individual cells. If one boiled parts of plants, individual cells were sometimes released and their separation from each other was amply demonstrated. Link tried this procedure, coupled with gentle pressure, in coloured beans, in the roots of several garden weeds and in other plants. He was able to isolate not only the cells of the parenchyma but also the elongated bast cells. For these, and a variety of other reasons, Link sided with Bernhardi and with his co-prizewinners Rudolphi and Treviranus in finding no visible pores in cell walls. Thinking it possible that Mirbel had mistaken starch grains for pores, Link isolated these grains and chemically characterized them. Rudolphi proposed that Mirbel had mistaken the air sacs for pores, and Sprengel, who will be considered presently, suggested that intracellular granules (Körner) might have been the source of error. But the question of communication between the cells remained, for it was assumed that the sap must somehow pass from one cell to another. Link's final opinion was that this took place through invisible pores as a kind of transpiration (Durchschwitzen). 'One shouldn't be surprised at this; in the animal body too, moisture very often passes through such invisible openings.'*

Mirbel did not frankly admit his error, but with the passage of time he did modify his position. In the *Elémens de physiologie végétale et de botanique*,[21] written in 1815, he still states that contiguous cells share a common wall and that this wall is riddled with pores or slits; but already in 1808 in *Exposition et défense de ma théorie*[22] the pores have become undiscernible: 'First of all, it can hardly be denied that the membrane in plants is riddled with countless undiscernible pores which facilitate the movement of fluid.'* And the movement of fluid has become very slow: 'Le tissu cellulaire ne reçoit les fluides et ne les transmet que très-lentement.' (The cellular tissue receives fluids and transmits them only very slowly.) The *Défense* contains both German and French texts, but Mirbel excuses himself for not being absolutely clear about the German criticisms of his work, for he knows no German himself and is unable to find anyone who is thoroughly conversant with the German language and who is also a botanist. Almost all German scientists at the time read, and often wrote, French; the reverse was much less common.

Hypothetical pores in cell walls were not, however, the only bone of contention. There was also the fundamental question of how new cells were formed. Link divided the immature plant into three components: 'das blasenförmige Gewebe' (*contextus vesiculosus*) (vesicular tissue); 'das fasrige Gewebe' (*contextus floccosus*) (fibrous tissue); and 'das fädige Gewebe' (*contextus filamentosus*) (filamentous tissue). He referred in the first chapter of the *Grundlehren* to Pitton de Tournefort who believed that the fibres of plants were composed of cells and communicated this view to the Académie des sciences in Paris in 1692. Link, despite the passage of more than a century, did not dissent from Tournefort's view. He was, moreover, convinced that new cells were formed in the intercellular spaces: 'Obviously new cellular tissue is formed between older cells. In the intercellular spaces, where later one sees simple passages, one can observe in young tissue dark streaks which appear to be composed of compressed material and which often reveal an extremely fine tangle of threads and other hardly recognizable components.'[23]* And to explain the mechanism of cell formation, Link fell back on vague notions that are reminiscent of some aspects of *Naturphilosophie*: 'Organic material is first formed from fluid stored in organic receptacles through the operation of a force that Blumenbach has appropriately called the creative urge, but whose laws are unknown to us.'* The shape and size of the formed elements were thought to be determined by the continued growth of the elementary structures generated by this creative urge. 'The formed elements grow by the insertion everywhere of the most minute, invisible particles. In this way new parts develop in the intercellular spaces of the older cells, and old parts differ from the new only in the number and size of the individual components.'[24]*

Mirbel's final reply to this was delivered in 1835, in his study of *Marchantia polymorpha*.[25] He begins by stating that thirty years had now elapsed since he first put forward the views that were so vigorously attacked. But the present work, he claims, is not so much a criticism of others as a piece of self-criticism. It dealt only with *Marchantia*, because he believed that more is to be gained by studying one organism in depth than by the superficial examination of a range of organisms. However, there is little new in this work, and it is surprising that Mirbel appears to have taken no cognizance of the progress in plant anatomy that had taken place since his initial contribution and the criticisms that were made of it. In *Marchantia* he still saw one cell generating another until an assemblage of 'utricules' is formed, and his final words on the generation of cells were: 'Très-certainement ce n'est pas par l'alliance d'utricules d'abord libres que le tissu cellulaire se produit.' ('Most assuredly it is not by the union of utricles that are initially free that cellular tissue is produced.') Sachs[26] quotes an obituary notice of Mirbel by Milne-Edwards, from which it appears that failing health in later life rendered him apathetic and unable to carry out his official duties. He was, in Sachs's view, the founder of microscopic plant anatomy in France; but Sachs cannot resist adding that all that had gone on in this field in France prior to Mirbel was nonetheless less important than what had been achieved in Germany.

There are two much-quoted authors whose works straddle the three Göttingen prize essays, one a few years before that event and one a few years after. Kurt Sprengel's *Anleitung zur Kenntnis der Gewächse*[27] (Introduction to the study of plants) appeared in 1802, and it is of interest because it seems to be the origin, or one of the origins, of the idea that new cells were formed from intracellular corpuscles or granules. Sprengel also considered that the vesicles of which cellular tissues were composed were sometimes filled with sap, and sometimes with air. 'By cellular tissue we mean a collection of closed vesicles that are bound together and enclosed by delicate membranes. They have different, although commonly rectangular, shapes and contain either sap or, often, simply air.'* As for the walls of the cells, he says: 'The cell walls are the most delicate membranes, most commonly without visible openings or pores.'* He does, however, see pores in some species of plant, for example in conifers, and he also sees intercellular spaces.

Sprengel's evidence for the development of cells from intracellular granules was obtained mainly from an examination of the seed-leaves of germinated bean seeds. In this material he noticed the presence of small granules and vesicles within the cells, and surmised, because of their appearance, that they were perhaps later transformed into cells. It is not absolutely clear that what Sprengel observed were starch grains (it is not easy to see how starch grains would be mistaken for vesicles); but, as mentioned before, Link nonetheless thought it likely that Mirbel had mistaken starch grains for pores in cell walls, and he rejected Sprengel's views on cell generation essentially because he believed that Sprengel's vesicles were also starch grains. In any case Sprengel appears to have accepted, or at least been silenced by, Link's criticisms, because in the second edition of his *Anleitung*,[28] which appeared in 1817, he no longer proposes that cells arise from granules; in fact he now makes scant mention of cell generation at all. The idea that cells did originate from starch grains was later resurrected in 1825 with Raspail's *Développement de la fécule.*

The second author, Dietrich Georg Kieser (who published after the award of the Göttingen prize), is of interest for three reasons: first, because although he was a German and a professor at Jena, the definitive statement of his position, the *Mémoire sur l'organisation des plantes*[29] was written in French; second, because he wrote a widely read textbook, in German, in 1815, the *Elemente der Phytotomie*;[30] and third, because he is mentioned by Dumortier, the discoverer of binary fission, as one of the principal proponents of the view that cells arise from subcellular granules. The *Mémoire sur l'organisation des plantes* was submitted as a prize essay for a competition organized by the Teylerian Society on the subject of plant fine structure, and especially on the tubes and vessels to be found in plants. Kieser's essay won the prize and became very well known. It begins with a detailed review of the work of Hooke, Malpighi, Grew, Leeuwenhoek, Du Hamel de Monceau, Hill, van Marum, Mustel, Hedwig, Senebier, Sprengel, Treviranus, Link, Rudolphi and Mirbel; and it is, in this respect, one of the most useful of nineteenth-century compendiums of botanical history. As for Kieser's own ideas on the organization of plants, he takes the view that 'Les plantes sont composées *en grande partie* de cellules, qui forment un tissu cellulaire.' ('Plants are composed *in large part* [that is, not entirely] of cells, which form a cellular tissue.') He believed that these cells arose from an infinity of tiny, transparent globules to be found in the sap. These globules were thought to enlarge, cohere and, as a result of external pressures, assume their final shape. In the *Elemente der Phytotomie*, he restated this view, and he laid emphasis on the conclusion that cells were indeed the fundamental subunits of plants. He regarded the cells of lower plants, which retained their original ellipsoidal shape, as the most primitive forms. In higher plants, however, the cells were forced by reciprocal pressures into a shape that he now defined more specifically as a dodecahedral rhombus. These pressures were said to obey strict mathematical laws, but how the forces were exerted and what mathematical laws they obeyed are not stated. Kieser's work was well known in France as well as in Germany, for his *Mémoire* was readily accessible to French-speaking scientists, and his *Elemente der Phytotomie* was much quoted by their German counterparts. But Dumortier, whose work will be discussed in detail at a later stage, had no hesitation in dismissing the idea that cells originated in any kind of subcellular corpuscle, an idea that he associates principally with Treviranus and Kieser.

The French contribution to these debates was characterized less by meticulous observation than by speculative inference; the German by a striking mixture of both. The Germans were undoubtedly right on a number of specific secondary issues; but neither side made much headway in their efforts to answer the fundamental question of how cells were generated. That had to await the advent of Dumortier.

CHAPTER 5

The Views of Some Standard German Textbooks

These thorough, and often massive, works not only reflect the state of knowledge at the time, they also reveal the striking disparity between what was known and agreed about the fine structure of plants and the rudimentary notions that histologists entertained about the composition of animal tissues. It is of interest, for example, to compare Heinrich Ernst Weber's *Allgemeine Anatomie des menschlichen Körpers*,[1] which appeared in 1830, with Franz J. F. Meyen's *Phytotomie*,[2] which appeared in the same year. Mention has already been made of Weber's detailed analysis of artefacts produced by refractive errors in the compound microscopes then available. Nonetheless, after criticizing numerous authors for falling prey to these errors, he accepted it as established that animal tissues were composed of six elements: (1) granules (Körnchen); (2) semi-fluid amorphous material; (3) material of cellular composition; (4) fibres (Fasern); (5) tubules (Röhrchen); and (6) leaflets (Blättchen). Bones are said to have numerous leaflets and fibres; epidermis, on closer inspection, to be composed of cells; the leaflets that one sees in tendons turn out to be fibres, and so do the leaflets in the fluid contents on the crystalline lens. Many fibres, such as are seen in nerves, muscles, liquid egg white and blood seem to be made of linear arrays of granules (Körnchen) or vesicles (Kügelchen), although this observation is to be treated with reserve, because the magnifying power of microscopes does not permit the resolution of such small objects,

> We therefore do not know whether, as Gallini, Platner and Ackermann have assumed, all the components of the body, including the smallest fibres and leaflets, consist of a porous substance, that is, one that is interrupted by interstices, so that porous or cellular material thus becomes the fundamental ground substance of animal tissues; or whether the smallest granules, fibres and leaflets are on the other hand composed of a homogeneous material which cannot be divided into smaller units but which is nonetheless interrupted by the shapes it assumes and by the presence of interstices.[3*]

Weber makes a distinction between 'Kügelchen' (vesicles) and 'Körnchen' (granules) and sees the latter almost everywhere: in the chyle, lymph, serum, pigment, milk, pus, bile and saliva. He gives a clear description of fat globules (Fettbläschen) (*vesiculae adiposae*) and blood corpuscles (Blutkügelchen) (*globuli sanguinis*). Reviewing the morphology of red blood cells in a range of different species, he makes the prescient technical point that in the case of human red cells the central spot or shadow may, as some authors say, be a cell nucleus, but it may also be an area of increased density or simply an optical artefact. This point will be discussed in some detail when the work of Prévost and Dumas (1821)[4] is considered in connection with the discovery of the cell nucleus. Naturally, Weber is more precise in the description of 'Körnchen' in body fluids, but he also has a section on 'Körnchen' in solid tissues. Here it is not too clear what he actually saw, since he describes these particles in solidified egg-white ('geronnene Eiweiss'). In the matter of cell formation, he reviews several options but vacillates:

> Similarly, we are far from having satisfactory observations to account for the progressive formation of small components of different shape during the development of the human embryo. Many anatomists merely surmise that all those small components are formed from amorphous material and from vesicles both of which are present from the very beginning in the material from which the embryo arises; that many vesicles become hollow and give rise to cells; that fibres arise from solid vesicles that have come together in linear arrays, and tubules from hollow vesicles assembled in linear arrays and then joined together; and so forth. Indeed one can envisage many possibilities.*

But all the possibilities envisaged by Weber ultimately presupposed the formation of particles or vesicles from some kind of non-cellular ground substance. The generation of cells from cells, to say nothing of binary fission, was not discussed.

Meyen's *Phytotomie* embodies a different order of microscopical anatomy altogether. This is not only because plant cells are larger than animal cells. As discussed earlier, the limitations of microscopy on animal tissues lay not in the resolution of existing microscopes, but in the fact that the preparation of animal tissues for microscopy lagged far behind what was possible with plants. Indeed, it was not until Purkyně and Valentin invented the microtome and Remak introduced hardening agents that satisfactory microscopical preparations of animal tissues became available. Meyen was a doctor of medicine and surgery, but the scientific work for which he is remembered is limited to the study of plants. *Phytotomie* is accompanied by an atlas containing fourteen copperplates of sections of plants. The illustrations in this atlas show precise pictures, which can hardly be faulted, of the cellular systems of plants. The generally accepted view that plants were composed of cells is not questioned, and Meyen is principally concerned with the accurate description of these cells and the structures to which they give rise. As might be expected from the title of the book, he gives little attention to the mechanism of cell formation or other physiological problems that phytotomy in itself was unlikely to illuminate. The section of the book entitled 'Das System der Zellen' deals mainly with variations in the morphology, including variations in the size, of cells, and the assumption is made that more specialized structures are produced by the elongation of cells or by their differentiation in other ways.

Meyen also classifies cells, according to their morphology, into six different classes: (1) 'die Kugel' (spherical); (2) 'das Ellipsoid' (ellipsoidal); (3) 'die Walze' (cylindrical); (4) 'das Prisma' (prismatic); (5) 'die Tafel' (tabular) and, (6) 'der Stern' (stellate). He thinks that regular shapes are determined by strict mathematical laws, but has his doubts about cells of irregular shape. Some intercalated cells have a crystalline appearance, but Meyen insists that, despite appearances, they are not crystals. Fibres he believes are formed from cells, and he refers to the particular family of cells involved as 'Faserzellern' (fibre cells). One of the most interesting features of the book is the author's realization that there was a clear-cut homology between lower plants, including those that are unicellular, and flowering plants. Cells were thought to be the fundamental subunits of both and differed mainly in their size or shape. Already at this stage Meyen was interested in the results that had been reported on silkweeds and added some observations of his own on this material. In particular, the fact that silkweeds could easily be disaggregated into single units led him to accept the view that cells were independent entities with imperforate walls.

By 1837, in Meyen's three-volume treatise, the *Neues System der Pflanzen-Physiologie*[5] the mode of cell formation has become a central issue. In Volume 2 he refers to Morren's observations on cell division in *Crucigena*. These were reported in 1830, and will be discussed in a later section dealing with protists. Meyen had no access to *Crucigena*, but in a related species, *Scenedesmus*, he saw the division of a mother cell into two or four components. This observation is followed by a reference to Dumortier's work, which is of interest in itself.

> In the year 1832, M. Dumortier observed the multiplication of cells by means of true cell division. As soon as the terminal cells of *Conferva aurea* have grown significantly longer than their neighbours, a partition wall is formed within them, so that from a single cell two are always formed. At a later date, this kind of cell multiplication by means of cell division was observed by C. Morren in *Closteria* and by H. Mohl in *Conferva glomerata*. Nowadays, the number of such observations has greatly increased.*

It is clear that Meyen attributes priority for this discovery to Dumortier. But Sachs,[6] in his strongly Germanophile history, credits Mohl with having given the first detailed description of cell division in multicellular plants and is lavish in his praise for the meticulousness of Mohl's work. Dumortier is not suppressed altogether in Sach's book, but he is dismissed in a single bald sentence which states that Dumortier observed cell division in 1832, which was three years earlier than the first of Mohl's publications on this subject. Sachs does not refer to Dumortier's original paper, but to the *Neues System* of Meyen who is perfectly clear in allocating priority to Dumortier. Regrettably, almost all German texts, following Sachs, have been loyal protagonists of Mohl.

Despite his accurate description of cell multiplication by binary fission, Meyen nonetheless believes that cells can be formed within mother cells, and he gives anthers as a particularly good example of this form of endogeny. He adds a precise, if speculative, discussion of the mechanisms that might be involved. On the other hand in *Marchantia*, the formation of spores is said to occur by means of cell division: 'The most striking demonstration is provided by spore formation in the genus *Marchantia*, where the evidence is compelling that cell multiplication does not take place within so-called mother cells, but by means of cell division as it does in other mosses and liverworts.'[7]* In the end Meyen compromises. He points out that binary fission is known to be characteristic of many lower forms, but many doubt that it occurs in higher plants, a view espoused notably by Ehrenberg in his book *The Infusoria.*[8] Meyen disagrees: 'Such views must, however, give way to the direct observations of recent times, and I believe that I can demonstrate in what follows that cell multiplication by binary fission is an occurrence that is more widespread among plants than it is among animals.'* Then there follows a review of the evidence in plants including the observations on liverworts, *closteria*, oscillatoria and confervae. Meyen describes at length the apparent mechanism by which a new partition wall divides the cell into two, 'allmähliche Einschnürung und Abschnürung' (gradual constriction and segmentation), but he does not yet entertain the possibility that binary fission might be at the heart of all cell multiplication.

Rudolph Wagner, a distinguished professor of physiology in Göttingen, published his *Lehrbuch der vergleichenden Anatomie*[9] (Textbook of comparative anatomy) in 1834–5 and his *Icones Physiologicae*[10] in 1839. His treatment of cells in the widely read *Lehrbuch* is very superficial, and all he has to say about their mode of formation and their fundamental role in the construction of the organism is the following: 'The cultivation of this interesting field (the development of the laws governing organic morphogenesis) has lain fallow since Leeuwenhoek, Malphighi and Haller, and it is only very recently that it has been taken up again. The great improvement in microscopes and the growing interest of younger observers will shortly throw new light on this subject.'* Wagner, too, has his reservations about the appearances so far given by the microscope and limits his histological descriptions to what is readily observable and generally accepted. He gives an extended description of red blood cells and offers the information that in vertebrates they are nucleated. The colourless nucleus is said to be insoluble in water but enclosed by a membrane that is water-soluble and red in colour. He describes in some detail the colourless corpuscles that he finds in lymph and chyle but makes no mention of the prior work of William Hewson.[11] Solid tissues are classified in a completely conventional way and in some of them granules or corpuscles (Körnchen) are seen. The cornea is alleged to be a simple homogeneous structure composed of fibres, leaflets and cells, and teeth are seen as similarly constituted, the leaflets being formed by the conjunction of fibres. Cartilage, when examined under the microscope, shows small, rounded and rectangular corpuscles. Muscle is composed of bundles separated by 'Zellgewebe' (cellular tissue), which here must also mean connective tissue, and nerves are simple tubes from which 'Kügelchen' and 'Klümpchen' (vesicles and blobs) can be pressed. This observation had been made previously by Valentin (Chapter 9), and it is probable that the spherical bodies (Kugel) observed by him were droplets of myelin.

The *Icones Physiologicae* which has a foreword dated 18 December 1838, that is, before the dissemination of Schwann's *Mikroskopische Untersuchungen*, is much more informative. It is an atlas with a commentary in both Latin and German, and its subtitle makes clear its intention: *Erläuterungstafeln zur Physiologie und Entwicklungsgeschichte* (Explanatory plates illustrating physiology and embryology). Although the plates are mainly anatomical, function has moved to the forefront of Wagner's scientific interests. The first twelve plates deal with the development of the egg and clearly show the 'Keimfleck' (*vesicula germinativa*: germinal area) first described by Purkyně (see Chapter 9). Plate XIII shows human red cells arranged in rouleaux. In the red cells of *Proteus anguineus* a nucleus is seen and the observation made that it is revealed more clearly if the cells are treated with water (haemolysed). The red cells of *Triton cristatus* also show a nucleus, but this cannot be seen in chick embryo erythrocytes, presumably because of their small size. Epithelium from the vagina is shown as a pavement of nucleated cells. The illustrations of muscle reveal bundles (Muskelbündel) but no unequivocal statement is made about cells in muscle.

Plate XXI reproduces Purkyně's illustration of a human brain showing ganglion cells (Ganglienkugeln). This illustration was taken from Purkyně's lecture at a meeting of the Society of German Naturalists and Doctors in Prague on 19 September 1837 and, as discussed later, constitutes an important piece of the evidence on which Purkyně's case for priority over Schwann is based. It is clear that in the year or two that preceded the appearance of Schwann's monograph, cells and their constituents had been observed in several animal tissues, and their function had come to be a centre of interest for histologists. But there is nothing in Wagner's atlas to suggest that he regarded these cells as the basic elements of the whole animal body; and, although he must have known that Purkyně, in his lecture, had drawn an analogy between plant and animal cells, Wagner gives no indication that there might be a general correspondence between the two.

Heusinger's *System der Histologie*,[12] which appeared in the year preceding the publication of Milne-Edwards's thesis, gives essentially no empirical information about the genesis of cells in animal tissues and does not seem much concerned with this question. It begins with a long review of the previous histological literature, and it is only in the discussion of cartilage that we are given an indication of Heusinger's ideas about cell formation: 'They [the cells] arise as soft droplets of plastic lymph [formative tissue]. If the surroundings permit it they take on a quite globular shape; if the surroundings do not permit this, they become discs or more polygonal bodies. They soon become harder and altogether acquire the appearance of cartilage. I have therefore called them chondroids.'* It is evident that in 1822 Heusinger believed that cells arose from non-cellular material, and he made no distinction between cells and globules.

Arnold's *Lehrbuch der Physiologie des Menschen*[13] (Text-book of the physiology of man) of 1836 has hardly moved further. As has been mentioned, Arnold adopts, without questioning it, the model proposed by Milne-Edwards,[14] and his illustrations show animal tissues composed entirely of vesicles that are uniform in size and differ only in their disposition. But Johannes Müller's *Handbuch der Physiologie des Menschen*,[15] which was published in sections between 1833 and 1838, marks a serious transition. It begins with a consideration of organic material (organische Materie). This is a rather vague concept, and Müller, perhaps also taking his cue from Milne-Edwards, states that it is often composed of rounded, microscopic 'molecules', although he is well aware that the 'molecules' differ in size. Some of Müller's 'molecules' were apparently large enough to be seen with a microscope. Studnička[16] thinks that in the blastoderm of the chick embryo Müller might actually have seen genuine cells, for, in this tissue, he describes an aggregate of rather large vesicles; but, in the same breath, he remarks that these vesicles are comparable to those that he finds in egg-yolk. However, in his monograph on the *Myxinidae*,[17] which bears the date 1835, Müller comes seriously to grips with the morphology and putative physiology of animal cells. In the first chapter, on the osteology of *Myxinidae*, he discusses vertebrates in general and extends his observations to *Ammocoetes* and *Petromyzon*. The comparative histology of all parts of the skeleton is considered, and it is in this section of the work that Müller remarks on the appearance of the cells that he sees in the *chorda dorsalis* (the notochord): 'The cells are irregular and heterogeneous, but to some extent resemble plant cells in that their walls seem to be closed on all sides and usually abut on each other in straight lines. In sections, therefore, the cells appear as irregular polygons.'*

This famous passage is accompanied by an illustration that leaves little doubt that Müller was looking at real cells, but nowhere does he see nuclei within these cells. It was Müller who drew Schwann's attention to this resemblance between cells in the *chorda dorsalis* and those of plants. The influence of Müller on Schwann, not only in the matter of histological methods but also in the transmission of ideas, can hardly be overestimated.

This review of the most widely read textbooks of the day makes it clear that before 1838 the scientific community had no inkling of the ubiquity of cells in living forms. It was generally agreed that plants were largely composed of cells, and cells had indeed been seen in several animal tissues; but no one had suggested in print that plant and animal cells were homologous. Nor was there any agreed opinion about how cells were generated. Binary fission had been described, but its occurrence, when noted at all, was thought to be an exceptional mode of cell multiplication limited to certain lower forms of plant life. Nothing like binary fission had yet been observed in animal cells. In seeking an explanation for the impact made by the contributions of Purkyně and Schwann, the confused state of contemporary knowledge about these matters must be borne in mind.

CHAPTER 6

Little Animals

Much has been written about Leeuwenhoek's momentous discovery of 'little animals', notably the classic monograph by Dobell[1] and, more recently, the biography by Shierbeck.[2] Leeuwenhoek's observation that there were living creatures that could be seen only with a microscope opened up a whole new world for biologists to explore and is the uncontested foundation stone of the modern science of microbiology. But more than a century elapsed before anyone suggested that there might be a connection between his microscopic 'animalcules' and the cells of higher life forms. Leeuwenhoek reported what he saw in a series of letters to the Royal Society, written originally in Dutch[3] and by no means elegant Dutch. He had no knowledge of classical languages, so the letters had first to be translated into Latin, and summaries of them, in English, were then published in the *Philosophical Transactions* of the Royal Society. *The Collected Letters of Antoni van Leeuwenhoek*,[4] which gives both the original Dutch text and an English version of it, remains the best resource for those who have little or no Dutch; but the commentators sometimes read more into the text than Leeuwenhoek's simple, and often colourful, descriptions themselves reveal.

The original account of Leeuwenhoek's discovery of conjugation in his microscopic animals was given in a letter that he communicated to the Royal Society on 10 December 1681. It was published in Robert Hooke's *Lectiones Cutlerianae*[5] which was a collection, begun in 1679, of lectures that had been given before the Royal Society. The English version bears the title: 'Some Microscopical observations made by Mr Anthony Leeuwenhoek, concerning the Globulous Particles in Liquors, and the Animals in *Semine Masculino Insectorum*'. Here is the key passage:

> I have lately examined Water, in which beaten Pepper was steeped, and found two sorts of animals for shape, and each of those sorts to contain greater and smaller kinds: The greater I supposed the elder, the less the younger: I conceive I saw in some of the greater sorts, the young ones in their Bodies; and when I observed two swimming joyned together, I conceived they were coupled.

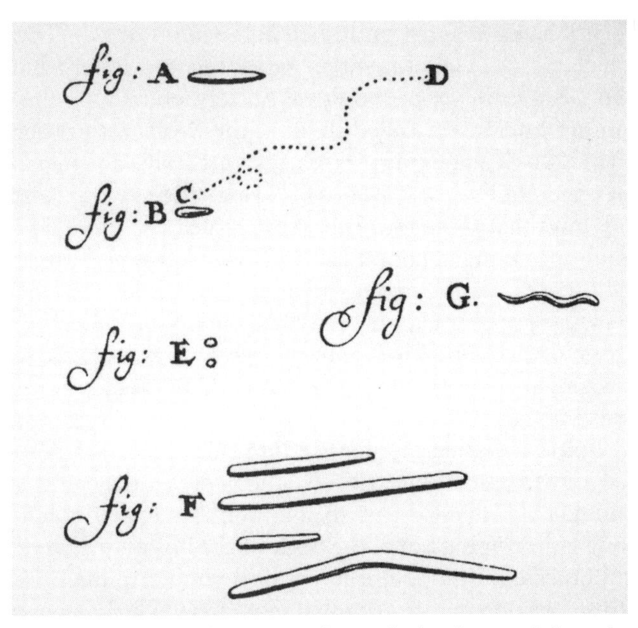

21. Leeuwenhoek's illustration of animalcules (bacteria) from the mouth

22. A facsimile of Leeuwenhoek's microscope, with an explanatory diagram

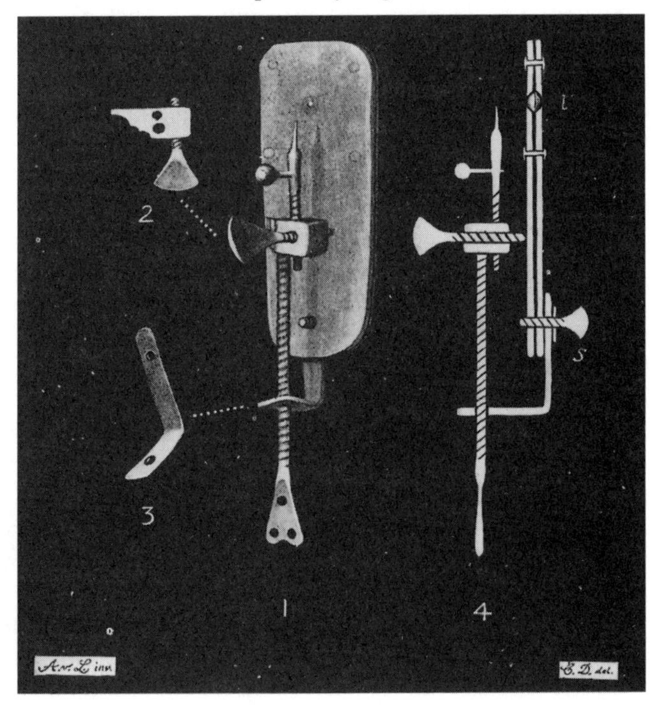

The summary of another letter, published in the *Philosophical Transactions* in 1694[6] has an even more elaborate title: 'An extract of a letter from Monsieur Anthony van Leeuwenhoek to the Royal Society, containing the History of the Generation of an Insect called by him, the Wolf, with observations on insects bred in Rain-Water, in Apples, Cheese, etc.' This abstract contains the following passage: 'In Rain-water I observed a small red worm, and two other kinds of very minute insects; of those of the larger size I judged that 30,000 together would not equal a course sand. These I observed for several days, and saw them copulate, the larger dragging the smaller through the water after them, swimming by the means of very small fins.' And in 1703, in a 'Part of a letter from Mr Anthony van Leeuwenhoek FRS concerning green Weeds growing in water, and some animalcula found about them'[7] we find: 'Among these [animalcules], I observed some which were much bigger than the rest and were coupled together, in which action they lay very still on the sides of the glass.' In this same letter Leeuwenhoek states that he is sure he has seen one animalcule giving birth to another: 'In *Fig. 9* you may see an Animalculum, b, h, coming out of the body of the bigger, which *Phenomenon* when I first perceived, I thought it might be a young Animalculum fastened by chance to an old one; but observing it more narrowly, I saw it was a *Partus* [offspring] . . .' This conclusion is reinforced on the following page: 'Another *Animalculum* that had brought forth two young ones, had her body laden with another sort of little creatures . . .' It is clear that over a period of more than twenty years, what started as an impression had become a firmly established observation. Leeuwenhoek believed not only that his little animals copulated, but that there were both males and females (*her* body laden, etc.), and that the latter carried their young within their bodies and eventually gave birth to them. This view was accepted by John Needham (1713–81), questioned by Charles Bonnet (1720–93) and finally dismissed by Horace de Saussure (1740–99) and Lazzaro Spallanzani (1729–99).

Abraham Trembley (1710–84), working in Geneva, appears to have been the first to describe binary fission in any organism.[8] In a letter[9] to the President of the Royal Society, dated 6 November 1744, he describes the division of several species of newly discovered microscopic fresh-water polyps. These were minute animals which clustered together and which Réaumur consequently called 'des polypes en bouquet.' They were 'capable of swimming about' but they did eventually affix themselves to solid surfaces. Trembley gives an unequivocal account of their mode of multiplication: 'The stem or Pedicle of a polypus that is yet single and which has but lately fixed itself, is at first short, but it lengthens itself in a little time. After that, the *Polypus* multiplies; that is to say, it divides or splits itself into two lengthwise.' He describes in detail the rounding up of the animal prior to division and the withdrawal of its processes, and then it '. . . gradually splits down through the Middle, that is, the middle of the head to the place where the posterior end joins the pedicle.' Finally the newly formed polypi open up, show their 'lips' and other distinguishing features, and 'may be looked upon as entirely

23. Abraham Trembley (1710–84)

formed.' 'They are, at first, less than the *Polypus* from which they are formed; but they grow to the same size in a very little time.' 'A *Polypus* is an hour, or there about, dividing itself.' It would be difficult to surpass this concise and accurate description.

In 1747 Trembley, who had now become a Fellow of the Royal Society, wrote another letter to its President.[10] In this he describes his technique for keeping large numbers of water creatures alive and the apparatus he has devised for this purpose. He doubts whether he would have discovered the manner of multiplication of these animals without this apparatus, and he offers the further information that observations on their behaviour are enhanced if their locomotion is reduced by cold temperature or by fasting. Another detailed description of binary fission is given, and summarized thus: 'It then by degrees puts on a round form, and presently after the little spherical body so formed, divides itself into two other like spherical bodies. These last in a few moments again insensibly open, they then lose their spherical form, and put on that of a bell, or of a *Polypus* as perfect and as compleat as that by the division of which it was formed.' Trembley observed this form of division in a number of other microscopic species: the colonial vorticellid *Epistylis anastatica*, *Carchesium* and *Zoothamnium*. In 1766, in a letter to Count Bentinck, cited by Baker,[11] he described binary fission in *Synedra.* Since binary fission was eventually shown to be the general mechanism underlying the multiplication of cells in all higher plant and animal forms, it is not unreasonable to regard Trembley's discovery as second in importance only to that of Leeuwenhoek himself.

In 1776, Lazzaro Spallanzani (1729–99) published in Modena a record of the observations he had made on both animals and plants (the *Opuscoli*),[12] and in 1780 he produced a more detailed account in two volumes (the *Dissertazioni*).[13] A French translation of Spallanzani's work by Jean Senebier

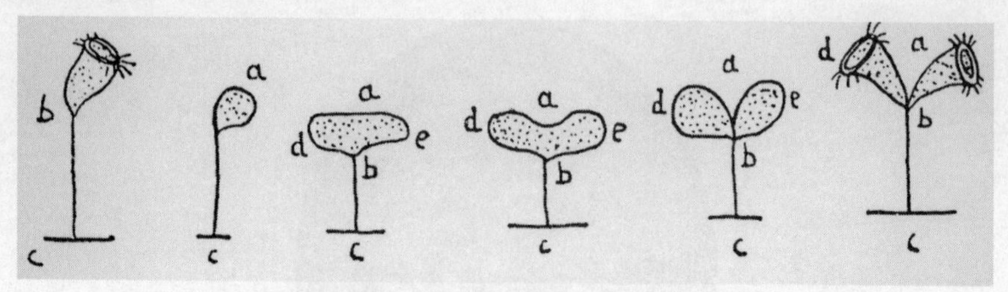

24. Trembley's illustration of binary fission in *Synedra*

appeared in 1786 bearing the title *Observations et expériences faites sur les animalcules des infusions*.[14] Chapter 9 of this work deals with the multiplication of these 'infusoria'. Spallanzani mentions that the animalcules are frequently seen coupled together, an appearance which, after Leeuwenhoek, many people regarded as copulation. Beccaria, who wrote to Spallanzani about the matter, advanced this view, but Spallanzani, under the influence of Trembley, had his doubts, implying that these ideas were interpretations read into the data by analogy with the behaviour of higher organisms. He recounts that he had received a letter on 27 January 1770 from Charles Bonnet (1720–93) who suggested that animalcules joined together in this way might not be copulating but might be dividing in the manner Trembley had described for the 'polypes à bouquet'. Spallanzani then produces a letter that he had already received from Horace de Saussure (1740–99) in which there is a detailed description of how the animalcules divide into two transversely. Saussure writes that the waist of the dumbell progressively narrows until the two animalcules are connected by no more than a filament, and it is at this

25. Lazzaro Spallanzani (1729–99)

stage that one may be misled into thinking they are copulating, particularly because of the vigorous movements that the two daughters make to be free of each other. These daughters are initially smaller than their parent, but quickly regain their normal size and may then again themselves divide. This repeated division and subdivision generates a population of increasing density. Saussure describes division in two species, one of which has an anterior hook, which it withdraws when it rounds up and prepares to divide. As in the case of the 'polypes à bouquet', an animal that has rounded up may undergo a rotatory movement. Saussure, like Trembley before him, is meticulous in his observations and Spallanzani agrees with what he says. In all, Spallanzani claims to have observed binary fission in fourteen species of infusioria and to have seen both transverse and longitudinal division. John Ellis, FRS (?1710–76) remained resistant to the idea of binary fission and advanced arguments in favour of copulation, particularly the 'copulatory movement' of the animalcules that were joined together, but Spallanzani dismisses Ellis's arguments and adheres to the view that the movements are not part of copulation but part of the division of the animalcules into two. The observations of Saussure and Spallanzani sounded the death-knell of the idea that primitive forms participated in a sex life analogous to our own. But the idea died hard.

In 1756 O. F. Müller in his celebrated book on the infusoria found in the sea and in rivers,[15] described binary fission in the microscopic green alga, *Vibrio lunula* and also in a fresh water flagellate. But it was not until 1830 that Morren realized that at least some of the infusoria were single cells and that the division that he saw in them was cell division. Charles Morren was at the University of Ghent, which was then French-speaking, and his papers, written in French, were submitted to the *Annales des sciences naturelles*. The 1830 paper[16] deals with a new kind of microscopic plant, which he names 'Crucigénie' (*Crucigenia)* and with an apparatus called a 'Microsoter' that he has designed to keep microscopic creatures alive. He first saw *Crucigenia* organisms in groups of four: 'When I first saw my little plant it had the shape of an extremely regular Maltese cross.'* But Morren recognized that the Maltese cross was not a single organism but was formed by the apposition of four cells that were separated by visible lines: '. . . their individuality does not reside in a collection of creatures but in each one of these creatures itself'.* He noted in one case a limb of the Maltese cross coming apart and forming in turn a perfect cross, from which he was able to conclude that the four arms of the cross were formed by two successive divisions.

Morren knew of the prior work of Trembley on polyps and acknowledged the relevance of Trembley's observations to his own. It is clear that he regarded the constituents of the Maltese cross as cells and does not hesitate to call them that. The description given to one of the figures in his paper is: 'Primary cells which divide into four and which, by barely perceptible gradations, turn into the different stages of young Crucigenias.'*

In 1836 Morren, again in the *Annales des sciences naturelles*, published a paper on *Closterium*.[17] He refers to the work of Dumortier on *Conferva aurea* and, like him, sees a midline partition that divides the *Closterium* into two. The cell first polarizes and then:

> . . . from the whole of the periphery in the middle of the *Closterium* where its diameter is narrowest, a circular plate had spread out towards the centre, separating the polarized masses of coloured particles like a transparent partition. . . . The partition is thus formed in the present instance as it is in the *Confervae* in accordance with the splendid observations of M. Dumortier*

The formation of the partition marks the beginning of cell division: 'After the middle partition is thus formed, the external junction becomes apparent as a black circular streak which exactly delimits the shared base of the two cones that make up the *Closterium*. This streak marks the *dehiscence*, which will take place in the plant later. . . .'*

Christian Gottfried Ehrenberg's book *Die Infusionsthierchen als vollkommene Organismen* (Infusoria as complete organisms)[18] is a magnificent tome, both from an aesthetic as well as an historical point of view. Ehrenberg was, at the time of its publication, an Extraordinarius in Berlin, but, a year later, he was made an Ordinarius in medicine there and remained attached to the University of Berlin for the rest of his academic life. As early as 1830 he had reported the division into two of *Actinophrys*,[19] and in 1832 longitudinal fission in *Euglena acus*.[20] But in the *Infusionsthierchen* (1838) he assembled morphological descriptions of some 120 families of infusoria from bacteria to *Paramecium* and *Polytoma*. The book, which was accompanied by an atlas containing a host of splendid coloured drawings, remained for decades the standard work on the taxonomy and morphology of these organisms; but, to a modern eye, it is a remarkable blend of meticulous observation and unbridled fiction. It is dedicated to Friedrich Wilhelm, Crown Prince of Prussia, and begins with an extensive historical review. Then there follows a section on the methods of collecting and maintaining these infusoria, which makes the point that they are not to be found in evil-smelling puddles. ('Die Infusorien findet man nicht in übelriechenden Pfützen.')

The rest of the book is a compilation of Ehrenberg's exhaustive observations on the numerous families of microscopic organisms that one finds in water. The classification that Ehrenberg offers is not one that would be at all accepted today. For example, in the First Family, which he classifies as Monads, he includes bacteria (Vibrio), *Uvella*, *Polytoma*, *Pandorina*, *Gonium* and *Chlamidomonas*. However, it is not in his taxonomy of infusoria that modern interest lies; it is in his views of their mode of multiplication. He has by now observed binary fission (Theilung) often enough for him to have no doubts about its existence. But he regards 'Theilung' as a special case and does not entertain the idea that it might be a universal mechanism. Indeed, he regards it as so exceptional that he classifies the organisms that multiply in this way as a separate family, to which he gives the name 'Theilmonade' (Monads that undergo division). They are to be distinguished from clustering monads (Traubenmonaden) by the fact that they show incomplete segmentation. 'The genus comprising monads that undergo division can be distinguished from clustering monads by the fact that when they divide, partitioning of the individual is incomplete.'* Ehrenberg admits that both longitudinal and transverse division are prominent: 'Voluntary transverse and longitudinal division are, however, very eye-catching.'* But, as a general rule, he believes the infusoria are to be divided into male and female categories and that each organism contains elements that correspond to the sexual organs. Thus: 'The reproductive apparatus of the Monads . . . consists of very numerous concatenated granules scattered throughout the whole body and a relatively large round glandular body which divides when the organism as a whole divides. This glandular structure is . . . obviously quite analogous to a masculine testis and the granules closely resemble eggs.'* Even in monads that undergo division, sexual organs comparable to those seen in the Trematodes are thought to be present. 'As far as the masculine element of the sexual system is concerned, the shapes that it assumes in no way militate against its identification; indeed in many species it is very clear.'* Ehrenberg sees chloroplasts but he imagines them to be 'green eggs' and the brownish particles that he sees at the same time he regards as eggs also. For him, the infusoria are hermaphrodites that combine within the one organism both male and female characters. 'The female part is present as coloured, uniform, very numerous granules, the eggs; the male part forms 1–2 rounded, usually very prominent, glands and separate contractile vesicles.'* Although all this is imaginary, it must be said that it was widely held at the time that binary fission occurred only exceptionally.

At this point it is of interest to turn to a remarkably prophetic work in which, apparently for the first time, the gap between primitive and higher animal forms was bridged. In 1805 Lorenz Oken (1779–1851), whose family name was actually Okenfuss, wrote the monograph entitled *Die Zeugung*[21] (procreation or generation) in which he advanced the view that all living forms were composed of 'infusoria'. In 1805 'infusoria' included, as Oken himself put it, all the organisms that grew in animal or plant infusions 'whether open to the air or covered, cold or warm'. They ranged from bacteria to complex protozoa, and there is little reason to doubt that for Oken the term simply meant the simplest and most primitive forms of life then known. In later works, written when more was known about the infusoria and a vesicular or globular structure was proposed for animal tissues, Oken abandoned the use of the word 'infusoria' and substituted 'Urbläschen' (primary or primitive vesicles). His original thesis was that all multicellular forms were merely aggregates of 'infusoria' that retained their individual existence, but cooperated to form the particular plant or animal that they constituted. He rejected spontaneous generation or *generatio equivoca* resulting from the coalescence

26. Lorenz Oken (1779–1851)

of organic material. The growth of living organisms in plant or animal infusions he believed to be simply the dissolution of multicellular forms into their basic components.

In considering the growth of a higher plant, Oken leans heavily on the observations of Trembley. He refers to the multiplication of microscopic polyps by binary fission and argues that growth of complex tissues must similarly be due to the multiplication of 'infusoria' or, later, 'Urbläschen'. These in his view undergo differentiation to form specialized organs. In the case of higher plants, he argues that it is their polarity that determines the elongation of the 'Urbläschen'; and in the spaces between them, vessels are formed that carry the sap. It is clear that Oken had an extraordinarily prescient view of how animals and higher plants are organized, and he was evidently convinced of the basic homology between unicellular and multicellular forms. Baker considered that Oken had little influence essentially because he produced no experimental evidence in support of his ideas. In this, I think Baker was mistaken. Oken was a man of enormous influence in his time. The son of an Alemanic farmer, he rose to be a professor at Jena and eventually filled a chair at the newly formed University of Zürich. He was an anatomist of high repute, and there are parts of the embryo that still bear his name, for example Oken's body (the primitive kidney) and Oken's canal (the embryonic precursor of the excretory duct of the testis). His adamant rejection of teleological explanations of form was well known, and he was noted for the anatomical thesis that the skull was simply a modified vertebra: 'Der ganze Mensch ist nur ein Wirbelbein.' (The whole of a human being is merely a vertebra.) He was the editor of an abstracting journal, *Isis*, and, as mentioned previously, the founder of the Gesellschaft deutscher Naturforscher und

Ärzte, which still confers a Lorenz-Oken medal. As a *Naturphilosoph* Oken had a European reputation. His *Lehrbuch der Naturphilosophie*[22] ran to three editions, the first appearing in 1809 and the last in 1843. It was translated into English and was widely read. In 1835 he wrote a popular exposition of natural history, the *Allgemeine Naturgeschichte für alle Stände*[23] (A general natural history for all levels). And in the matter of the cellular composition of animal and plant tissues, he disputed priority with Schwann. The reason for the apparent indifference to the ideas put forward in *Die Zeugung* was not so much the lack of experimental evidence as the fact that these ideas were far too early for the scientific readership of the time. If *Die Zeugung* had been written some twenty-five years later, the book would surely have made a substantial impact.

CHAPTER 7

Dumortier and Mohl

Barthélemy Charles Dumortier (1797–1878) was the victim of a serious historiographical injustice. Both before and after the appearance of Sach's history[1] of botany, the German scientific literature attributed the discovery of binary fission in multicellular organisms to Hugo Mohl (later von Mohl), a now legendary professor of botany at Tübingen. There is no doubt that Mohl did see cell division and described it accurately, but he did not see it first, and it is open to question whether he saw it independently.

In a retrospective collection of articles presented to his father in 1845,[2] Mohl reproduces the famous 1837 paper from the *Allgemeine botanische Zeitung*[3] in which he describes cell division in multicellular plants, mainly in *Conferva glomerata*. He makes a point of saying that this paper was a revised version of a doctoral thesis presented in 1835. The data first appeared in *Flora* in 1837[4] and bore the title 'Ueber die Vermehrung der Pflanzenzellen durch Theilung'. (On the multiplication of plant cells by means of division.) No date earlier than 1835 can be found for Mohl's observations and no date earlier than 1837 for their publication. Yet the paper in which Dumortier described cell division in *Conferva aurea* appeared in 1832[5] and was actually presented to the Académie des sciences in Paris by Cuvier in 1829. Dumortier was aware of his priority and referred to it in a paper that appeared in 1837,[6] the year in which Mohl first published his findings. It is of interest to examine how it came about that Dumortier was largely overlooked (except in his own country, and then mainly as a politician), while Mohl was generally credited with having made the discovery.

Sachs's *History*, which ends in 1860, mentions Dumortier but in an entirely unsatisfactory way. Mohl is extolled as the prince of phytotomists and several pages are devoted to the discussion of his numerous contributions to botany. His 1837 papers on cell division are singled out for special praise because, for the first time, it is alleged, the whole process is described in detail and accurately. Dumortier's 1832 paper on *Conferva aurea* is bracketed with Morren's 1836 paper on *Closterium*, and both are found wanting in detail. No mention is made of Morren's 1830 paper on *Crucigenia* or of the fact that Morren's observations were made on protista,[7] while Dumortier's were made on a multicellular organism. As mentioned earlier, Sachs does not cite Dumortier's original paper, but merely the reference to Dumortier in Meyen's *Neues*

27. Barthélemy Dumortier (1797–1878)

System.[8] But Meyen is absolutely clear about the primacy of Dumortier's work, and specifically states that both Morren's observations of *Closteria* and Mohl's on *Conferva glomerata* came later.

It is difficult to understand why Sachs did not refer to Dumortier's original paper rather than to Meyen's abstract of it, and why he should have condemned Dumortier's work for want of detail. As will be shown, Dumortier's paper is extremely detailed, and there are reasons for believing that Sachs should have known about it. It seems very unlikely in any case that Mohl did not know about it, for one would then have to assume that his choice of *Conferva* as experimental material was mere coincidence, or that silkweeds were a commonplace object of study, a proposition which does not seem to be reflected in the literature of the early 1830s. In any case, Mohl does not mention Dumortier in his 1837 paper,[9] nor does he in the selected works reissued and discussed in 1845,[10] although he does refer to Nägeli's work on *Conferva glomerata*, published in 1844.[11] Dumortier's elision from the literature, except on occasion as a name and a date, was not confined to the serried ranks of German followers of Mohl and Sachs. Even François Duchesnau in his *Genèse de la théorie cellulaire* (1987),[12] which is written in French and presents, in many respects, a Gallic point of view, refers only to Dumortier's later work on the embryology of gastropod molluscs.[13] His 1832 paper on cell division in *Conferva aurea* is not mentioned at all, either in the text or in the bibliography.

Dumortier was a Belgian, became actively engaged in politics in mid-life and eventually rose to ministerial rank. As was almost universal for Belgian scholars at that time, he wrote in French, but the work on cell multiplication by binary fission, although deposited earlier with the Académie des sciences, was published in the *Transactions of the Imperial Leopoldino-Caroline Academy*,[14] a journal that was widely read not only in Germany but throughout Europe. It is for this reason that I think it unlikely that Mohl, or, for that matter, Sachs could have been completely unaware of Dumortier's work. The paper bore the all-embracing title 'Recherches sur la structure comparée et le développement des animaux et des végétaux.' (Investigation of the comparative structure and the development of animals and plants.) The author notes that he is a Membre de l'Académie and could therefore not have been altogether unknown to other European scientists. Here are some extracts from the paper, which will permit the reader to judge the cogency of Sachs's assertion that the description of cell division given by Dumortier lacked detail.

> The development of the confervae is as simple as their structure. It is brought about by the addition of new cells to old, and this addition always takes place at the tip [of the filament]. The terminal cell elongates to a greater extent than those below it (see plate 10, fig. 15a), and half-way down the cell in the internal fluid an extension of the inner cell wall is produced in such a way that it divides the cell into two parts. The lower of these remains stationary (see plate 10, fig. 15b) whereas the upper, now terminal, again produces a new internal partition and so on. There is no way to determine whether this partition at the mid-point is, in principle, single or double, but it is certainly true that later on it appears to be double in conjugated filaments (see plate 11, fig. 34e), and when two cells come apart of themselves, each of them is closed at its extremity. This can be easily demonstrated in the confervae when they have matured or when the cellular tissue has been frozen, for in that condition the individual cells still enclose the fluids that they previously contained, which would not be possible if they were not sealed by a membrane.
>
> The observed production of a midline partition in the confervae seems to us to provide a perfectly clear explanation of the origin and development of cells, which has hitherto remained unexplained. . . .*

Dumortier makes three points. First, it is the terminal cell in the filament that elongates and then forms a midline partition which divides the cell into two. Second, this dividing wall can be presumed to be double because the individual cell retains its contents when it is separated from the filament. The double nature of this wall can be observed directly in conjugating confervae at a later stage. Third, the division of the cell into two provides a rational explanation for the origin of cells and their subsequent development. The first two points are adequately illustrated in the plates. It is difficult to construct a clearer and more concise account of the basic features of cell division than that given by Dumortier.

Baker suggests that Dumortier might have been mistaken in supposing that cell division takes place only in the terminal cell of the filament. Mohl's

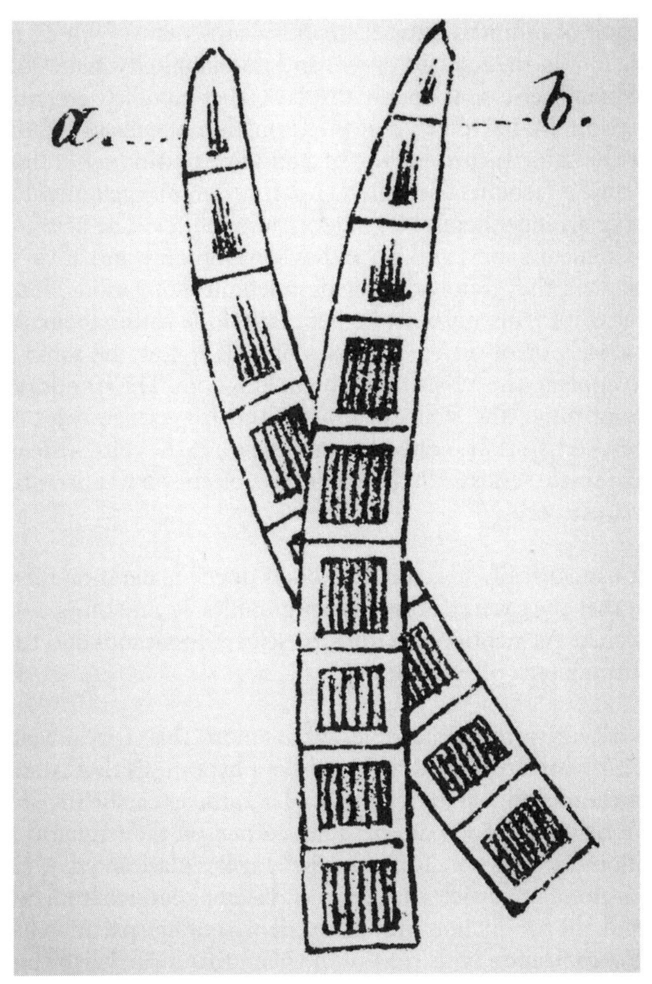

28. Dumortier's illustration of binary fission and partition formation in *Conferva aurea*

illustration of cell division in *Conferva glomerata* certainly involves a cell that is not terminal.[15] But Nägeli in 1844[16] confirmed that in this species cell division took place only in the terminal cell, except where branching occurred. The process was then also seen at the branch-points. It is possible that the cell illustrated by Mohl was at a branch-point, for he emphasizes the role of cell division in the formation of new branches. In any case, although Mohl examined six species of confervae in addition to *Conferva glomerata*, on which most of his observations were made, these did not include *Conferva aurea* with which Dumortier worked.

This was an important point for Dumortier. His whole conception of plant growth rested on the linearity that he observed in *Conferva aurea*:

> The growth of infinitely simple creatures presents us with an anatomical structure that is already laid down, and this simplicity makes observation less uncertain because it openly displays what complex creatures hide in their interior. We have seen that the formation of cells in the confervae is brought about by the production of a mid-line partition; but this occurs in a strictly linear fashion. The cells do not form lateral agglomerations or join together or arrange themselves into organic centres. The new cells are laid down in a linear series, and since they develop only and always at the tip of the filament they follow the law of indefinite elongation. Here again the confervae display openly what higher plants hide within them. All production of vessels or of fibres, all series of cells follow the same law which seems to apply to the whole of the plant kingdom. The fronds of algae, the thalluses of fungi, the stalks of mosses and Jungermanniales display the same characteristics. It is only that in these cases the cells, instead of developing in a single series as in the confervae, often show substantial connections between series.*

Dumortier categorically rejected all models of cell generation that rested on the notion that they were formed from granules or any other form of subcellular particle. As mentioned earlier, he selects Treviranus and Kieser as the principal proponents of such models.

> Mr Kieser's view is thus untenable. What's more, this view, as well as the one advanced by Mr Treviranus, is based on a hypothesis that is anything but proven – the idea that granules or globular particles can be transformed into cells. We believe that we can be quite certain on the strength of our own observations that this transformation never takes place and that starch grains as well as globular particles are entirely different entities from cells. On the other hand, the production within the cell of a midline partition fits in so well with the organization of the rest of the plant that it can hardly be denied.*

The second part of the paper, dealing principally with animal cells, has a completely different character. Whereas the large cells of silkweeds could be observed directly, the cells of animal tissues were probably inaccessible to the techniques that Dumortier had at his disposal in 1828. For this reason what he says about animal cells is largely speculative and has a touch of *Naturphilosophie* about it. Impressed by the fact that he saw cell division in the terminal cell of the silkweed filament, Dumortier proposed that all growth in plants was 'centrifugal', that is, that it involved only the extremities of the plant, the growing tips of stems, roots and branches. Growth in the animal, however, was thought to be 'centripetal'. By this Dumortier meant that the generation of new cells took place in the interior of the body and that these cells, on migration to the exterior, became static. However, it is notable that the paper of 1832[17] contains no evidence of an experimental kind, either microscopic or manipulative, concerning animal cells.

While there is thus no doubt that Dumortier has priority over Mohl in the discovery of cell division in multicellular plants, and even in demonstrating the duplex nature of the dividing wall, of which Mohl later makes much, Hughes, in his *History of Cytology*,[18] claims that cell division in silkweeds was observed much earlier by Vaucher. I doubt this. Vaucher's book, *Histoire des conferves d'eau douce* (Sweet-water silkweeds) was published in 1803 in Geneva.[19] It is a beautifully illustrated book that is mainly concerned with the taxonomy of these plants and with their modes of sexual reproduction (conjugation, fertilization, sporulation, germination). Vegetative growth receives scant attention, but Vaucher does observe the growth of the filament from the germinated spore and increased partitioning of the hypha as elongation proceeds. Here is the relevant passage:

> At almost the same moment and on the same day, or at least in the same week, all the seeds of *conferva iugalis* (I had several thousands) opened up at one end just like the two cotyledons of a seed from which an embryo is being formed; and from the bottom of the opening a green sack emerged, very small at first, but which soon elongated to such an extent that it was several times as long as the globule itself. In the interior of the sack, spirals soon appeared. They were accompanied by their bright dots as in a fully developed plant. The tube itself showed partitions, first one, then two, then a larger number. Finally the offspring broke away from its seed and floated alone on the liquid. Then, apart from its size and its two ends which were still pointed, it completely resembled the plant that gave birth to it.*

However, Vaucher nowhere describes the actual formation of the dividing wall, nor does he say that it divides the cell into two. He merely notes that the number of partitions increases as the plant grows. If he had seen cell division, it is difficult to believe that he would have omitted to say where he saw it, either in the terminal cell of the filament, as Dumortier insisted was the case in *Conferva aurea*, or elsewhere. It is therefore difficult to avoid the conclusion that priority for the discovery of cell division in a multicellular organism remains with Dumortier.

However in the observation that the partitions in the silkweed filament were duplex in nature, Vaucher did anticipate Dumortier:

> As far as the internal partitions that divide the plant are concerned, they are, like the tube itself, formed by a very fine transparent membrane. Although they appear single, I have reason to believe that they are double; for I have often seen the tube separate into two, or into three, or even into as many pieces as the number of compartments that the tube itself originally contained. And since these separated compartments do not lose their contents but retain the green material and the spirals that they originally enclosed, it must be presumed that they are shut tight. Otherwise the phenomenon that I have observed could not have taken

> place. The tubes of silkweed with which we are here concerned can therefore be considered not as single plants but rather as aggregates of a large number of plants. On this view, each compartment in the tube is itself a plant which does not communicate with the others contained in the same tube. One compartment can be apposed to another or separated from it. Each has its own envelope, its spirals, its particles, in a word, everything that constitutes a plant, and, as we shall soon see, it can reproduce itself.*

Vaucher thus uses the disaggregation of the plant into sealed units as evidence both for the independence of these units and also for the partition that separates them being double. This decisively antedates similar conclusions drawn by G. R. Treviranus, Moldenhawer, Link and Dutrochet from the various disaggregation procedures that they used for higher plants. G. R. Treviranus acknowledges this in his book.[20]

Dumortier adhered in a general way to the belief that there was a uniform structure for both animal and plant tissues, but he did not think that their mode of multiplication was necessarily the same. He believed that the cells of some animal tissues were generated by binary fission, as in the case of *Conferva aurea*, but that in others progeny was produced by a different mechanism. Dumortier maintained his opposition to models in which cells were formed from subcellular particles, but he seems to have been influenced by current theories of cell formation as expounded by Milne-Edwards, Turpin and Brisseau-Mirbel. And it may be that he was especially influenced by the observations of Armand de Quatrefages.

In 1834 Quatrefages published a paper[21] on the embryology of planorbes and other limnetic animals. In particular he concentrated on the development of the eggs of *Limnaeus ovalis*; and this was the experimental material chosen by Dumortier for most of the work described in the latter's 1837 paper on the embryology of gastropod molluscs.[22] The key passages in Quatrefages' paper are as follows:

> *1st day.* In the vicinity of the big ends of the egg, one perceives three or four rather large oval globules (1/97 mm). These are at first separated, but after a few hours they form irregular groups. Examining them closely, one sees that they enclose within them other much smaller globules (globulins).*
>
> *2nd day.* The number of globules increases, but they do not look any different. They form a kind of irregularly scalloped cake, a little less transparent in the centre than at the edges; there one sees that the globules are already piled on each other, while on top of these they are merely stuck together.*
>
> And finally: we know that the characteristic property of cells, especially at this stage of embryonic life, is to act as a kind of womb for other cells (globulins) which are formed within their interior.*

Dumortier's observations on *Limnaeus ovalis* are much more precise than those of Quatrefages. Six hours after the egg has been laid he sees in it 'un hile muqueux et diaphane, qui est la vésicule de Purkyně' (a mucous, diaphanous hilum, which is Purkyně's vesicle). Then, on the second day, he sees an appearance that could be interpreted as cell division, but he does not say so: 'The embryonic globule has grown considerably and is now twice its original size. . . . The hilum, for its part, has elongated and seems to consist of two diaphanous globules which soon separate and detach themselves from each other.'* On the third day, he sees the blastula but thinks that it represents only a series of superficial furrows on the surface of the globule. 'A notable change has taken place in the embryonic globule which has taken on a shape altogether different from what it showed yesterday. Its periphery is divided into five shallow lobes; the centre of the globule is more diaphanous than its periphery . . . and this globule now shows irregular facets on its surface.'*

The appearance of the embryo at this stage was said to resemble that described by Prévost and Dumas[23] for the frog embryo. But it clearly does not occur to Dumortier that he is seeing segmentation. In a summary of his findings he reverts to the notion that cells are generated inside other cells: 'Then an important phenomen takes place: inside the primary cells one begins to see secondary cells which increase in number each day and end up destroying the primary cells. Only their walls persist and turn into a network of small vessels.'* And in the conclusion to his paper, he comes perilously close to proposing the generation of cells from non-cellular material, an idea that in his observations on *Conferva aurea* he flatly rejected: 'Thus it is the surface of the embryonic globule that forms the first general tissue, just as it is the surface of the deposits of which it is composed that becomes the first internal cellular tissue. Thus, the original transformation of fluids that can be organized into tissues takes place by the solidification of their surfaces.'*

In the end Dumortier vacillates between the cell division that he has seen in silkweeds and popular models of animal cell generation that he has not seen:

> We have seen in the course of embryonic development two modes of tissue development, that found in the liver where the cellular tissue grows by means of mid-line projections like those that I was the first to describe in plants, and that found in dermo-muscular tissue which is propagated by the centripetal production of canaliculi which form the infiltrating pad that one observes there. This completely overturns the notion that animal tissues are formed only in one way, as advocated by Bordeaux, Meckel and others, and one is forced to recognize the plurality of mechanisms by which animal tissues are formed, as proposed by Bichat and his school.*

Dumortier is important not only because of his discovery of binary fission in multicellular organisms, which was epochal enough, but also because,

29. Hugo von Mohl (1805–72)

perhaps more than any other scientist working at that time, he illustrates the enormous gulf between the observations that had been made on plant cells and the fantasies that engaged the minds of those who worked with animal cells. As pointed out previously, this was not principally due to the limitations of the microscope. After all, within a year or two of Dumortier's 1837 paper, Purkyně and Schwann were making their observations on genuine animal cells; and although their microscopes were probably better than the one used by Dumortier, it was the development, and widespread adoption, of techniques that permitted animal tissues to be adequately prepared for microscopy that made all the difference.

The fact that he was decisively anticipated by Dumortier in a matter of great significance does not detract from Mohl's exceptional standing as an experimental botanist. Mohl's work always strikes one as painstaking, even pedantic, and the range of his contributions to botany is remarkable. His famous 1837 paper on cell division[24] begins with a rather polemical review of previous work. Like Dumortier, he dismisses the notion that cells could be formed from starch grains, chloroplasts or other particles. He regards these proposals as 'grösstentheils auf rein aus der Luft gegriffenen Vermuthungen beruhenden Angaben' (assertions that rest for the most part on totally unfounded speculations). But the main target of his criticism is Mirbel. He rejects Mirbel's model of cell formation and substitutes the mechanism that he has himself observed in *Conferva glomerata*. As mentioned earlier, he appears to have seen binary fission mainly at the branch points, and he concludes that it has an essential role in the formation of new branches: 'The branches of the stem always arise at the upper end of its cells and are separated from these cells by a partition. . . . At its junction with the stem, the cell about to form a branch develops a circular narrowing that projects into its interior. At this site it partitions the green cell contents and thus produces a circular dividing wall perforated in the middle.'*

A year after the appearance of his 1837 papers on binary fission, he confirmed, without mentioning them, the observations of Vaucher and of Dumortier on the duplex nature of the dividing wall; but his was probably the first accurate description of the splitting of this partition into two lamellae.[25] In 1839 he described the division of spore mother cells in *Anthocerus*,[26] but made no reference to the work of Adolphe Brogniart, who, as early as 1827, had given an accurate, if vague, description of the formation of pollen grains in the mother cells of *Cobaea scandens*,[27] or to similar observations that Mirbel had made on *Marchantia* in 1835.[28] By the time Mohl published the selection of his work in 1845[29] he had observed cell division in many species of silkweed in addition to *Conferva glomerata*: *Conferva fracta*, *cristata*, *rupestris*, *aegagropila*, *prolifera* and *Hutchinsiae*. He also saw it in *Callithamnion Rothii*, *repens* and *roseum*, in *Ectocarpus littoralis*, in *Draparnaldia plumosa* and *tenuis*, in several types of *Chaetophora* and in other species such as *Zygnema*.

The conclusion Mohl reaches at the end is that cell division by means of a partition is not altogether rare among confervae. Baker sees this conclusion as 'charming diffidence', but this, it seems to me, misinterprets both the text and the character of the man. In fact, Mohl's papers do not at all give the impression of diffidence, and, after Schleiden's dramatic paper of 1838[30] appeared, Mohl largely adopted Schleiden's position on cell generation and, with some reservations, adhered to it until after 1845. He does not remark on the similarity between the formation of spores or pollen grains and the binary fission that he had seen in so many lower plants. Indeed, he believed for many years that in higher plants the normal process of cell multiplication was the formation of new cells *de novo* from non-cellular material. What he had seen in the silkweeds and elsewhere he long regarded as a variation which, in his own words, was not all that infrequent in primitive organisms.

Mohl made many other observations that are of special interest to botanists. He was the first, or so he claimed, to describe the formation of vessels from rows of closed cells; he derived the cell walls of plants from a thickening of the cell membrane; he provided further evidence for the cellular origin of fibres. But perhaps his most important general contribution was the decisive distinction that he made between the sap of the plant and the internal contents of its cells. These had tended to be confused in earlier writings, but after Mohl's papers of 1844[31] and 1846,[32] the question was settled. Some historians have credited Mohl with the discovery of protoplasm; but this is an error even for the word itself, to say nothing of the substance.

In 1835 Félix Dujardin, who eventually became professor of zoology and botany at Rennes, had already published an accurate description of the

30. Félix Dujardin (1801–60)

cellular contents, and had called the material of which they were composed the 'sarcode'.[33] Dujardin began his investigations by attempting to confirm the notions that Ehrenberg had propagated concerning the intracellular digestive apparatus of infusoria; but he failed to do so. He proposed instead that the behaviour of these organisms was to be explained by the properties of the 'sarcode': 'I propose to give this name to what other observers have called a living jelly, this glutinous, diaphanous substance, insoluble in water, that contracts into globular lumps, sticks to dissecting needles, and can be drawn out like mucus. It is to be found in all lower animals interposed between other structural elements.'*

Note that Dujardin does not claim to be the first to notice this substance; he merely defines it and gives it a name. Unsurprisingly Mohl does not refer to him nor does he mention the fact that Valentin, of whom more will be said later, had already proposed the word 'parenchym' to describe the same material.[34] Valentin's paper dealt with the composition of nerves in which he consistently found a 'tough granular ground substance': 'They [the contents of the nerve] are always composed of a granular *Parenchym*, in which greyish-red, tiny particles are permeated by a softish, tough, transparent cellular material that is cohesive. In the middle or close to it there is a round or oval *nucleus* which is composed of a limiting membrane and a very bright interior.'* However, Mohl does refer to the now little-known Hartig and the less well-known term that the latter coined to describe the cell contents: the 'ptychode'. After describing the cell membrane Mohl writes: 'I found this inner cellular structure, which, for reasons that I shall later explain, I propose to call *primitive sac*, *utriculus primordialis*, in a similarly perfect condition in a range of dicotyledonous plants . . . e.g. *Sambucus Ebulus*, *Ficus Carica*, *Pinus* *sylvestris*, etc. . . . What I have said permits the conclusion that Hartig knew about this primitive sac and described it as *ptychode*.'*

But in his 1846 paper[35] on the movement of sap, Mohl is more specific and gives the *utriculus primordialis* a definitive name:

> As I have already mentioned, wherever cells are formed, this tough fluid precedes the first solid structures that indicate the presence of future cells. Moreover, we must assume that this substance furnishes the material for the formation of the nucleus and of the primitive sac, not only because these structures are closely apposed to it, but also because they react to iodine in the same way. We must assume also that the organization of this substance is the process that inaugurates the formation of new cells. It therefore seems justifiable for me to propose a name that refers to its physiological function: I propose the word *protoplasma*.*

All this clearly reveals the influence of Schleiden. Indeed, Mohl specifically states that Schleiden uses the word 'Schleim' (perhaps now translatable only as 'slime') for the substance he now names 'protoplasma'. Actually the word 'protoplasma' to describe the basic ground substance of cells appears to have been first used by Purkyně in a lecture that he gave in 1839, but this lecture, which appeared in print in 1840,[36] was given to the Silesian Society for National Culture, and it is reasonable to assume that Mohl did not know of it. Like Dujardin, Mohl himself made no claim to have discovered the substance to which he gave the name protoplasm. There were, in addition to Dujardin and his 'sarcode', prior descriptions of this substance by several observers including Jones[37] and Kützing.[38] Jones described the protoplasm in *Hydra* as 'semi-fluid albuminious matter'; and Kützing adopted Valentin's word 'parenchym' for it. Kützing believed, however, that the 'parenchym' was only present in a small minority of algae and evidently regarded it as a specialized layer that lined the inner surface of the cell wall. He gave it the additional name 'Amylidzelle', but Nägeli, who had also observed a 'Schleimschicht' (layer of slime) attached to the inner surface of the cell wall, objected strongly to Kützing's term,[39] and it did not gain general currency. But the name proposed by Purkyně and re-invented, presumably innocently, by Mohl was at once accepted by German scientists, and not long after by almost everyone else; whereas Dujardin's 'sarcode' was soon lost in obscurity.

CHAPTER 8

The Discovery of the Cell Nucleus

The discovery of the cell nucleus is only marginally less important than the discovery of the cell itself, for without an understanding of the structure and function of the nucleus, no coherent explanation of the mechanisms that transmit hereditary characters is possible. Most textbooks attribute this discovery to Leeuwenhoek. The attribution is probable but not certain. The relevant passage occurs in another letter from Leeuwenhoek to Robert Hooke.[1] It is dated 3 March 1682 and the autograph manuscript survives in the archives of the Royal Society. Here is the English translation given:

> Thus I came to observe the blood of a cod and of a salmon, which I also found to consist of hardly anything but oval figures, and however closely I tried to observe these, I could not make out of what parts these oval particles consisted, for it seemed to me that some of them enclosed in a small space a little round body or globule, and at some distance from this body there was round the globule a clear ring and round the ring again a slowly shadowing contour, forming the circumference of a globule and, as well as I could, represented in Fig. 5. And in others I saw 3, 4, 5, 6, nay as many as 8 globules, much smaller than the first globule, and though I observed this blood of the above-mentioned fishes without further delay than two minutes, and though they were fully alive, except the ray, I yet intend, because it is winter now and cold, to pursue my observations in summer and during warm weather.

It is apparent that Leeuwenhoek's description is tentative, but the editors of his complete correspondence[2] state categorically in the margin: 'Discovery of cellular nucleus'. There is no doubt that the Fig. 5 to which Leeuwenhoek refers clearly shows a drawing of the red cell of a fish, and it does contain a central body that we would now unhesitatingly call a nucleus. However, Leeuwenhoek also sees additional smaller globules in the red cell which the editors of his correspondence consider to be the vacuoles that form around the nucleus when the blood is heated to 43–45°C. But Leeuwenhoek remarks that his observations were made in winter and that it was cold, so it is not clear how a temperature of 43–45°C would be attained. The heat generated by oblique light passing through Leeuwenhoek's simple lens can hardly have

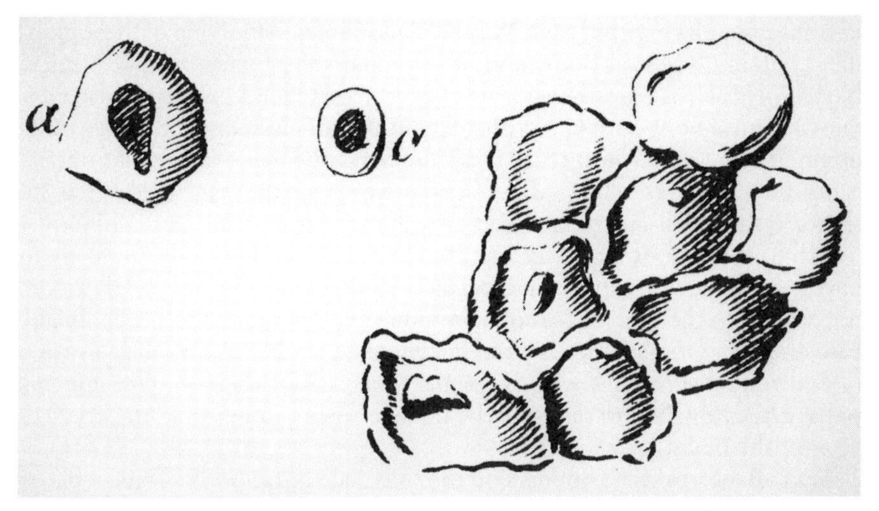

31. Fontana's illustrations of epithelial cells from the skin of an eel. A patch of cells is shown, a single cell at a higher magnification (a) and a red cell (c) for comparison. The epithelial cells show recognizable nuclei

raised the temperature of the specimen to this extent. Moreover, Leeuwenhoek in the same paper described 'globules' in the liver, and the editors say that they do not know what these might have been. In fact, as we have noted, Leeuwenhoek saw globules everywhere. It is not in any case clear that the nucleus would be seen in cells replete with blood pigment, although it can be seen in unfixed and unstained fish erythrocytes with a modern microscope. It is, of course, possible that the blood examined by Leeuwenhoek contained immature cells or cells that had lost some of their pigment. However, when he examined frog erythrocytes, which are much bigger than those of the fish, he failed to observe an internal globule, and merely described a central shadowing in the cell.

A survey of red cells from different species, written almost a century and a half later by Prévost and Dumas,[3] raises further doubts about Leeuwenhoek's observation. In an excellent plate which illustrates these red cells, a central body or shadow is shown in all of them, including the red cells of man where no nucleus exists. Only those of the goat, which are the smallest illustrated, show no central body. The central body or shadow is prominent in the huge oval red cells of the salamander and the frog, but its presence in anucleate human red cells raises the possibility that early observers might have been looking at an optical illusion.

Similar doubts arise in connection with the observations of Fontana.[4] Fontana illustrates a mass of globules that form the gluten of the skin of eels. They appear to be filled with very small particles. In a figure that shows these same globules in a preparation which is slightly dried out, a small body is seen in a different place within each globule. An enlargement of one of these globules contains a central body, and, for comparison of their relative sizes, a blood corpuscle is placed beside it. However, this blood corpuscle also contains a central body and, unfortunately, we are not given any indication of its origin. It is not elliptical and can therefore hardly be that of an eel. If Fontana is using a human red blood cell for comparison, then the central body that he shows in this cell is an artefact. Be that as it may, his drawing of an epidermal cell from the skin of an eel carries conviction. Epidermal cells are likely to have been flattened, especially if partially desiccated, and they are therefore not subject to the optical illusion commonly seen in human red cells. In any case, the illustration shown is immediately recognizable to a modern eye as a typical epithelial cell. If these arguments are accepted, then Fontana appears to have been the first to describe the nucleus in cells other than the erythrocytes of the blood.

Franz Bauer, or as he appears on the title page of his book, Francis Bauer Esq. FRS, was born in Feldsberg which was then in Austria, but later became Valtice in Czechoslovakia. He and his younger brother Ferdinand acquired formidable reputations as botanical draughtsmen and both had close connections with botanists in England. Franz Bauer began his illustrations of European plants in 1791 and completed them in 1798. These drawings were well known to Fellows of the Royal Society at the time and are now one of the most famous of such collections. They were not, however, published until 1830–38, the preface to the collection being dated December 1837.[5] There is no doubt that Robert Brown knew about them when he wrote the famous paper in which the nucleus is first given its modern name, for in that paper Brown refers to the earlier observations of Bauer.

The illustration in which Bauer draws attention to the cell nucleus is taken from a sketch made in 1802. It shows anatomical views of the stigma and stigmatic surface of *Bletia Tankervilliae*. The text accompanying the illustrations contains the following passages:

> The stigma is discharging the matter it contains; and the plexus lining the stigmatic canal is converted into a mass of loose separable oblong bodies, having their free extremities pointing upwards.
>
> Some of the same bodies magnified 200 times. They are perfectly transparent, and appear to be cellules of an oblong or fusiform figure, with one, two or three granular, more opaque greenish yellow specks, looking like young seeds of an Orchis in the midst of loose reticulated testa.

Bauer does not elaborate on the 'greenish yellow specks' that he observes, but his drawing shows eight cells of which six clearly have a single nucleus, one has two nuclei and one three nuclei. No cell is without its nucleus. Bauer also shows nuclei within cells in a longitudinal section of the surface of the upper part of the stigmatic canal, and in a transverse section of the dense

32. Robert Brown (1773–1858)

mucous substance lining the canal. The nuclei are clearly shown in cells magnified 100 and 200 times and in cells examined under water. As far as can be assessed from his drawings, Bauer regarded the nucleus not as an occasional, but as a regular feature of the cells he examined, and he recorded its presence in more than one tissue. However, the text accompanying the drawings is sparse.

Robert Brown's paper was published in 1833, but read to the Linnean Society on 1 November and 15 November 1831.[6] He was at the time custodian of the botanical collections of the British Museum and later became the president of the Linnean Society. Brown makes much more of the nucleus than Bauer appears to have done, but, contrary to the assertion by Baker and many subsequent historians, he does not claim that it is present in all cells. Quite the contrary, as the following passage shows:

> This nucleus of the cell is not confined to Orchideae, but is equally manifest in many other Monocotyledenous families; and I have even found it, hitherto however in very few cases, in the epidermis of Dicotyledenous plants; though in this primary division it may perhaps be said to exist in the early stages of development of the pollen. Among Monocotyledones the orders in which it is more remarkable are Liliaceae, Hemerocallidae, Asphodeleae, Irideae, and Commelineae.

Brown does, however, stress the point that the cells of *Bletia Tankervilliae*, illustrated by Bauer, were the only ones in which he found more than one nucleus. I think it likely that Brown was convinced that the nucleus had an important function, but he did not envisage that it had a universal,

33. Rudolf Wagner (1805–64)

indispensable function. Of course, he did not claim to have discovered it. Apart from the work of Bauer, he refers to prior descriptions of this organelle by Meyen, Purkyně and Brogniart. So we are left with the name 'nucleus' that he chose for it.

The history of names is not entirely rational. Brown first suggests the term 'nucleus', as an alternative to 'areola', in the following sentence: 'This areola, or nucleus as perhaps it might be termed, is not confined to the epidermis, being also found not only in the pubescence of the surface, particularly when jointed, as in Cypripedium, but in many cases in the parenchyma or internal cells of the tissue, especially when these are free from deposition of granular matter.' Throughout his paper he uses the terms 'areola' and 'nucleus' interchangeably. Why 'nucleus' survived and 'areola' did not is barely explicable. Perhaps it was because the Latin meanings of 'nucleus' carried the connotation of something solid, whereas those of 'areola' implied an open space. But at the time that Brown wrote his paper it had not yet been decided whether the nucleus was solid or vesicular. Nonetheless, in English and French writing, the term 'nucleus' was rapidly adopted, and in German it was often used as an alternative to 'Kern' (kernel). Purkyně's discovery of the *vesicula germinativa* in the hen's egg was made in 1825 and thus antedates Brown, but not Bauer. This discovery and its relevance to the theory of the cell nucleus are discussed in some detail in the following chapter.

Finally we have to consider the delineation of the nucleolus within the cell nucleus. Although this may well have been seen by some of the microscopists who made observations on the nucleus during the 1830s, the first accurate description of this organelle seems to have been that of Rudolph Wagner (1835)[7] whose *Lehrbuch der vergleichenden Anatomie* and *Icones Physiologicae* were discussed in Chapter 5. Examining the Graafian follicle of the sheep, Wagner noticed within the *vesicula germinativa* of the egg 'a

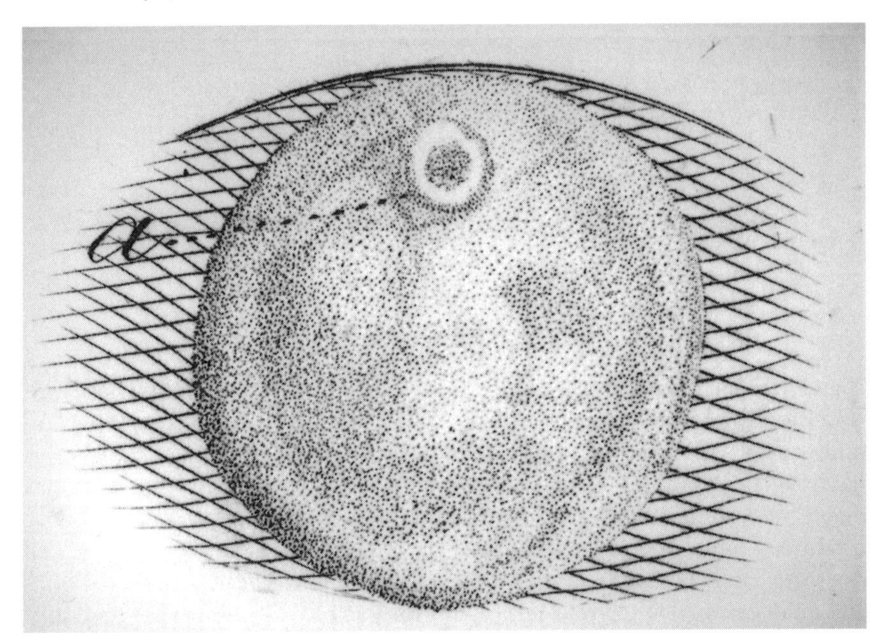

34. Wagner's illustration of a nucleolus within a nucleus

yellowish refractile, dark spot'; 'once I saw instead of a single spot, two smaller ones lying close together'.* 'My attention was drawn to this spot, because I also noticed it in other classes of animal. I am still doubtful whether it is always to be found in vertebrates, but it will be obvious to any observer in *Phalangium opilio*.'* The nucleolus is indeed clearly illustrated in a number of figures in the plates that accompany this paper.

Still vacillating about the constancy of the 'Fleck', Wagner proposes a name for it: 'This spot, which I should like to believe is constant, at least in mammals, I call the germinative spot (*macula germinativa*).'* In an appendix, he lists the numerous classes of animal in which he has seen it during the previous year. As to its function, Wagner surmised that it was the origin or first stage in the development of the germinal vesicle and eventually the germinal layer: 'The first sign of an embryo is what I have called the germinal spot . . . I have clearly observed the germinal layer arising from the germinal spot.'* In regarding the nucleolus as the centre of cell formation, Wagner anticipates one of the main features of the erroneous model proposed by Schwann.

With the detection of the nucleus and nucleolus the optical limits of the microscopes then available were reached. No further analysis of structural detail was possible, and apart from an occasional vague conjecture, no progress was made in the analysis of function. That had to await the discovery of chromosomes.

CHAPTER 9

The Cradle of Histology

Jan Evangelista Purkyně (1787–1869) is an example of a scientist whose reputation far from adequately honours his work, and for reasons that are quite other than scientific. Here I deliberately use the Czech version of his name, rather than the German, for one needs to explain why a man of this magnitude, both in his personal work and in the work of his students, should have been largely overlooked in conventional historical accounts of the cell doctrine, or, if not quite overlooked, then dismissed with a couple of lines and an occasional eponym. The reasons for this relative neglect are no doubt complex, but it seems to me that several distinct strands can be detected in the complicated web of circumstances that brought it about. To begin with, there was the uncomfortable position of a Czech nationalist in a cultural environment dominated in the nineteenth century by the German language. Then there was the civic rivalry between Berlin and Breslau, compounded by the personal rivalry between Johannes Müller and his school in Berlin and Purkyně and his pupils in Breslau. This was reflected in the selective quotation of scientific work by Müller and his followers. Müller came to occupy a dominant position in the realm of German physiology and most subsequent German writers adhered to the version of events that he propagated. Among the most decisive factors was the barely explicable, but widespread, enthusiasm for the doctrines propounded by Schwann in his monograph,[1] coupled with the modest attitude of Purkyně to the publication of his own work.

Since their defeat in the battle of the White Mountain in 1620, the Bohemian people had been inundated by an increasing tide of Germanization. During the next two centuries, Czech speakers had been progressively reduced to menial positions, and the distinction between the upper class and lower class was essentially a matter of whether one spoke German or Czech. The University of Prague, which was founded by Charles IV in 1348 and was originally open to Czechs, Germans and Poles, had, by the time Mozart made his celebrated journey to that city in 1787, become a completely German-speaking institution. However, the end of the eighteenth century saw the revival of a Czech nationalist movement and in the *Kulturkampf* of the nineteenth century an educated middle class emerged that spoke Czech.

35. Jan Evangelista Purkyně (1787–1869)

Purkinje, as he is usually spelt in the German literature, was the product of this *Kulturkampf.* He was born in Libochovice (Libochowitz), on the Eder, and received his early education as a choirboy in a Piarist foundation in Mikulov (Nikolsburg) in Moravia. After a period teaching in Litomysl (Leitomyschl), he left the Order and completed a secular education in medicine and philosophy at the University of Prague, where he graduated as a doctor of medicine in 1819. From 1819 onward he was an assistant and prosector in the anatomical institute of the university and in 1823 accepted the chair of physiology and pathology in the University of Breslau (or Wroclaw as it was called by the Poles then, and again, now).

Breslau was an outpost of German hegemony in Silesia and, although it was politically and culturally a German city, it had a large Polish subpopulation, and, as a commercial centre at the crossroads between East and West, harboured many other national minorities. The University of Breslau was founded in 1811, one year after the opening of the University of Berlin, and rapidly acquired a distinguished faculty in both the sciences and the humanities. The civic and cultural growth of the provincial city was accompanied by a great deal of local pride which included pride in the intellectual achievements of the newly founded university. The proximity of Poland and the presence of the Polish underclass engendered a fierce Teutonic patriotism in the German-speaking population, but this did not militate against rivalry between the university of Breslau and that of Berlin. Purkyně, as the holder of a chair in Breslau and the founder of the first institute of physiology in

Germany, thus found himself in the mainstream of German intellectual ambitions, and there can have been little room for the Czech roots to which he was so closely attached. When, in 1850, he returned to Prague as Professor of Anatomy and Physiology, he became increasingly involved in the Czech national revival and began to write popular and scientific works in the Czech language. This was no doubt appreciated by the Czechs, but meant, for practical purposes, that these works could not be read by anyone else.

The fact that, in his later years, Purkyně was seen by the Germans as essentially a Czech cannot have helped to maintain his scientific reputation in Germany. It was not until the twentieth century that serious efforts were made to restore his position as a major contributor to the science of microscopic anatomy, and one of the pioneers, if not the principal pioneer, of the cell theory. It is not surprising that these revisions of textbook history should have been made by Czechs or under official Czech patronage. Two of them are of special importance and form the factual basis of much of this chapter. In 1927, F. K. Studnička, then at the University of Brno (Brünn), published a detailed study[2] of the contributions of Purkyně and his school to histology generally and to the discovery of animal cells in particular. This appeared in the proceedings of the Moravian Society for the Study of Natural Sciences, and remains the definitive work on this subject, even if it is perhaps in places more than a little partisan. The other readily available source of information about Purkyně is a symposium held in Prague in 1969 to mark the hundredth anniversary of his death. This symposium, edited by V. Kruta, was sponsored by the Czechoslovak Academy of Science.[3]

Although Meyen, who was an experienced microscopist, still had reservations in 1836[4] about the fidelity of compound microscopes, it does seem that, during the 1830s, and especially the late 1830s, improvements in optical instruments as well as improvements in the skill of those who used them, did produce a rapid and marked change in the quality of microscopic observations. For his experiments on the hen's egg, which resulted in the discovery of the germinal vesicle, Purkyně used only a hand lens. This work[5] was Purkyně's contribution to a *Festschrift* that the Breslau medical faculty prepared in 1825 for the fiftieth anniversary of the graduation of Johann Friedrich Blumenbach, a professor at Göttingen who was something of a Nestor among naturalists. In the section headed 'De evolutione vesiculae germinativae (Keimbläschen)' (On the development of the vesicula germinativa [germinal vesicle]), the following passages occur: 'Then there is seen on its internal surface a diaphanous vesicle that projects slightly and is surrounded by a thin halo of white rounded material which in the mature egg forms a hillock. . . . The consistency of the vesicle is indeed very delicate, so much so that in the smaller eggs it bursts at the slightest touch, much like a bubble.'* It is clear that Purkyně not only noticed the presence of the germinal area but, in attempting to isolate it, revealed its delicate vesicular structure. He believed, however, that this vesicle disappeared as the egg matured. In the section entitled *De evanescentia vesiculae germinativae* (On the transitory nature of the germinal vesicle), he writes: 'In the first place, if you examine the small scar after the yolk has been taken away from its support, you will nowhere find the vesicle that has previously been described within the scar on the egg.'* It was at the suggestion of von Baer and Coste, who will be discussed later, that the *vesicula germinativa* of Purkyně was named the *vesicula Purkinjii*.

In the introduction to his work *Über Entwickelungsgeschichte der Thiere* (On the embryology of animals), which was published in 1829, von Baer paid Purkyně an unusual compliment in declaring that he himself had little to say because Purkyně had virtually exhausted the subject.[6] Purkyně, like many others after him, regarded the germinal vesicle as a whole cell and not as a cell nucleus. At no time did he draw an analogy between this vesicle and the nucleus that he saw in many other animal cells. In 1834, almost a decade after Purkyně had discovered the *vesicula germinativa*, a Polish student of his by the name of Adolph Bernhardt submitted a doctoral thesis in which there is a description of a structure analogous to Purkyně's vesicle *within* a mammalian ovum. The notion that the vesicle might correspond to a cell nucleus could then hardly be avoided.

In 1832 Purkyně acquired a new achromatic compound microscope, made by Simon Plössl of Vienna, and with this instrument began a systematic study of the microanatomy of animal tissues. Müller in Berlin received a Plössl microscope at about the same time and embarked on a similar programme. Valentin, the most eminent of Purkyně's pupils, records that in 1836 a new microscope made by Pistor and Schiek of Berlin arrived in Purkyně's laboratory and added to the precision of the observations being made. Purkyně's attitude to publication was atypical, not to say extraordinary. Whereas Müller's protégés, for example Schwann, Henle and Virchow, produced books and monographs that attracted a great deal of attention and were accompanied by a fanfare from the reviewing journals, the prior discoveries of Purkyně's group were recorded mainly in doctoral theses or in lectures and brief reports. In some cases Purkyně did not even put his own name on the publications that were generated by the doctoral theses, even though his input was often decisive. It was apparently enough that an author was known to be a member of the Breslau school for the association with Purkyně himself to be acknowledged.

Studnička[7] has studied the doctoral theses of Purkyně's pupils and lists the following among the important contributions that they made to the theory of the cell. In 1833 Alphons Wendt, in upholding a dissertation on human skin, *De epidermide humana*, reported that it had a granular structure ('textura granulosa'), and his illustrations show the presence of distinguishable granules in the deeper layers. Within these granules there were still smaller granules and it seems clear that he was observing cells with nuclei. Wendt later described such granules in many, if not all, animal tissues. Carolus Deutsch, this time together with Purkyně, described corpuscles in bone (Knochen-Körperchen), a discovery that Purkyně had himself made earlier in cartilage.

In this work a new technique was used in which sections were cut from decalcified bone. Henle, in his textbook of 1841,[8] states that 'a new era in the study of bone was initiated by the introduction of the decalcification technique invented by Purkyně under whose guidance Deutsch wrote his dissertation'. It is thus clear that the Berlin school was well aware of the contributions from Breslau even when these were recorded in the form of theses or dissertations. The 'Körperchen' seen by Deutsch in bone did not suggest plant cells to him, but seemed rather to resemble 'infusoria': 'bestiolis quibusdam infusoriis haud absimilia' (not unlike certain infusoria). This is an echo of the views of Oken, with which Deutsch may well have been familiar. In the case of Isacus Raschkow, however, a specific comparison is made between animal and plant cells. In his inaugural dissertation, *Meletemata circa mammalium dentium evolutionem* (Studies on the development of mammalian teeth), submitted to the medical faculty in Breslau in 1835, Raschkow describes for the first time the development of dentine, enamel and cement in the tooth and notes that in the epithelium of the dental papilla there are 'Körnchen' that closely resemble the cells in the parenchyma of plants: 'parenchyma plantarum cellulis simillimum'.

There are two important points to be made about Raschkow's observation. First, Purkyně claimed that, from his earliest years as a histologist, he had harboured the notion that there was a parallelism between animal and plant cells. This idea he purported to have discussed with Valentin who incorporated it into a prize essay submitted to the Institut de France in 1834. More will be said about this essay presently, but it was not, in any case, unknown, for in addition to winning the prize, it won the praise of Alexander von Humboldt. An abstract of its contents appears in the first volume of the *Repertorium* founded by Valentin in 1836.[9] The second point of interest is the use of the word 'Körnchen'. Purkyně and his pupils generally described cells as 'Körnchen', but whereas Purkyně himself adhered pretty faithfully to this term, one also finds, especially in the work of his pupils, the words 'Körperchen', 'Kügelchen' and 'Zellen'. Studnička makes much of the fact that Purkyně chose, out of preference, the word 'Körnchen', whereas Schwann, of course, uses 'Zellen'. According to Studnička the word 'Zellen' implied an essentially empty interior and thus drew attention to the cell wall, whereas the word 'Körnchen' (a small grain or granule) drew attention to the interior of the cell.

It is difficult to establish the exact connotations of 'Zelle' and 'Körnchen' for scientists in the first half of the nineteenth century, but it seems to me that the distinction made by Studnička is rather strained. Dujardin had already given the name 'sarcode' to the semi-solid contents of the cell in 1835, and Schwann, after all, gives the nucleus an all-important role in cell generation. Moreover, Valentin uses the word 'Zelle' to describe compact semi-solid 'Körnchen', as for example the 'Epidermiszellen von Proteus anguineus' (epidermal cells of *Proteus anguineus*), and he uses the word 'Kern' to describe the nucleus within the 'Körnchen' or 'Zelle'. Moreover,

36. Gabriel Gustav Valentin (1810–83)

Raschkow mentions that if you apply pressure to the 'Körnchen' (which in the Latin he calls *cellulae*) they burst and release a fluid resembling lymph. Of course, the word *cellula* in classical Latin connotes an empty interior into which things can be put, and the word 'Korn' connotes a solid body from which life can spring, but whether these etymological nuances persisted into the scientific literature of the nineteenth century seems doubtful. In any case, Schwann quotes the work of Raschkow and Purkyně in the full knowledge that the 'Körnchen' they described were the 'Zellen' that he was interested in. In 1836 yet another dissertation was presented to the faculty in Breslau, this time by Mauritius Meckauer. It was entitled *De penitiori cartilaginum structura* (On the internal structure of cartilage). Meckauer provided an accurate description of cartilage cells, which he calls *acini* and of the nuclei that they contain, which he calls *acini centrales.* Cartilage cells had been described before, but this was apparently the first time that their nuclei were depicted. Meckauer also introduced the words *substantia fundamentalis* to describe the intercellular cartilaginous ground substance.

The work of Gabriel Gustav Valentin (1810–83) falls into a different category from that of Purkyně's other pupils, not only because of its much greater scope, but also because Valentin, although setting out as a pupil of Purkyně's, attained a position of some independence while he was still in Breslau and should be classified as a collaborator rather than as a pupil, as he appears in most textbooks. In his *Handbuch der Entwicklungsgeschichte* (Embryology) of 1835[10] Valentin describes 'Körnchen' or Kügelchen' in many tissues, for example ovary, spleen, bone and the pigment layer of the retina; and he observes that they are larger in amphibia and fish than in birds and mammals. He often refers to Purkyně, but usually says of a particular observation that it was made by 'Purkinje und ich' (Purkyně and me). He also remarks that the cartilaginous tissue of the labyrinth is composed of 'a beautiful, six-sided trabecula that looks almost like cellular plant tissue, and in it there are small round granules'.* In the matter of correspondence between the composition of plant and animal tissues, Valentin disputed priority with Schwann, but in the arguments that he adduced in favour of his own case he does not mention the Paris prize essay. The original manuscript contained close to a thousand quarto pages, and the Académie des sciences asked that for publication a reduced version be produced. More than three years elapsed before this was completed and the second manuscript, which itself was some 300 quarto pages, bears the date January 1838. It nonetheless remained unpublished, although Purkyně, who cited it as evidence of his own and Valentin's priority, still expected in 1840 that it would be so. This, astonishingly enough, did not occur until 1963, when an abridgement and commentary by Erich Hintzsche appeared in the Bernese Transactions of the Society for the History of Medicine and Biology.[11] Valentin's essay bears the title: *Histogeniae plantarum atque animalium inter se comparatae specimen* (A comparative study of histogenesis in plants and animals). In his monograph, Schwann refers to Valentin's *Handbuch der Entwickelungsgeschichte* and makes disparaging remarks about mere morphological similarities between animal and plant cells; but he makes no reference to the prize essay. Although it had still not been published when Schwann's monograph appeared, it seems unlikely that Schwann had not heard of it, given that it was well enough known to have engaged the admiration of von Humboldt who sought to induce the Académie des sciences to publish it. Hintzsche considers that Valentin's failure to use the essay to support his priority claim is evidence that even when he wrote the second version of it he did not yet accept the universality of the correspondence between animal and plant cells that Schwann's monograph advocated. It is also possible that Valentin did not cite the essay because it had not yet been published and because he knew that the earlier version never would be. It is, in any case, clear that the idea that there was homology between *some* plant cells and *some* animal cells long antedates Schwann's *Mikroskopische Untersuchungen*; and even if Valentin did not present his observations in the form of a dramatic generalization, as Schwann did, it is difficult to avoid the conclusion that his claim to priority was at least to some extent justified.

Valentin's observations on peripheral nerves, the central nervous system and the choroid plexus are of special interest. He saw 'grosse Kugeln' (large spherical objects) at many different sites in the nervous system including that of invertebrates. These were obviously ganglion cells, as the following passage indicates: 'Each of these spherical objects is encased in a more or less obvious cellular covering and contains its own mass of parenchyma, an independent nucleus or kernel, and within the latter a rounded transparent secondary nucleus.'* He says much the same thing about the conjunctiva:

> It consists of closely packed, rounded, rhomboidal or square cells (note that he uses the word 'Zellen') whose boundaries are made up of simple, thread-like lines. In every cell, without exception, there is a somewhat darker and more compact nucleus which is either round or elongated. It is usually to be found in the middle of the cell and is composed of finely granular material. Within it there is a sharply rounded body which thus constitutes a kind of nucleus within a nucleus.*

In choosing the word nucleus (apparently for the first time in animal cell biology)[12] Valentin *ipso facto* draws a parallel with plant cells, since this was already the accepted term for this structure in plants. It is also clear that he is describing what we now know to be the nucleolus, and in his *Repertorium* of the year 1839[13] he introduces this word. 'Nucleolus', of course, remains the standard term for the 'Keimfleck' or *macula germinativa* that Wagner first discovered.

In discussing the large cells in the central nervous system, Valentin does not omit to mention Purkyně's prior discovery of their presence in the cerebellum (Purkyně's cells). Valentin's remarks on the cells in the choroid plexus again emphasize the similarity between plant and animal cells. He says explicity that each cell 'contains within it a centrally placed, dark, round kernel, a formation that reminds one of the nucleus that one finds in the plant kingdom.'* Concerning the formation and function of the nucleus, Valentin's views vacillated, and could not, in any case, have been correct at that time. In his *Handbuch* of 1835[14] he advances the view that the cell nuclei in the choroid of the eye arose by precipitation, a notion that antedates one of the themes later proposed by Schwann. Valentin at first considered that the cell could arise by cavitation of the nucleus and subsequent expansion of the nuclear cavity. ('Dass Zellen durch Hohlwerden von Kernen entstehen können.') Schwann dismisses this view and states that cells are never hollow nuclei. In a commissioned article that he wrote in 1839 for Wagner's *Handwörterbuch der Physiologie*,[15] which was published in 1842, Valentin maintained that blood cells and others free in suspension could multiply by binary fission. He claimed that this division also involved the nucleus, which could then generate two or more daughter nuclei. On the other hand he believed that for most fixed cells, and he gives cartilage cells as an example, endogenous generation of daughters within a mother cell was the usual mechanism responsible for cell multiplication. He did not, however, regard the nucleus as an indispensable cell component, and he classified epithelia into those that contained nuclei and those that did not.

After the widespread dissemination of Schwann's monograph, Valentin began to waver and largely adopted Schwann's views on cell formation, thus indicating that, although sometimes obstinate and always touchy, he was not immune to waves of fashion. One has the impression that he was a brilliant pupil and a highly productive collaborator, but that he lacked the assured stability of Purkyně. In 1835 there was a precipitate deterioration in the

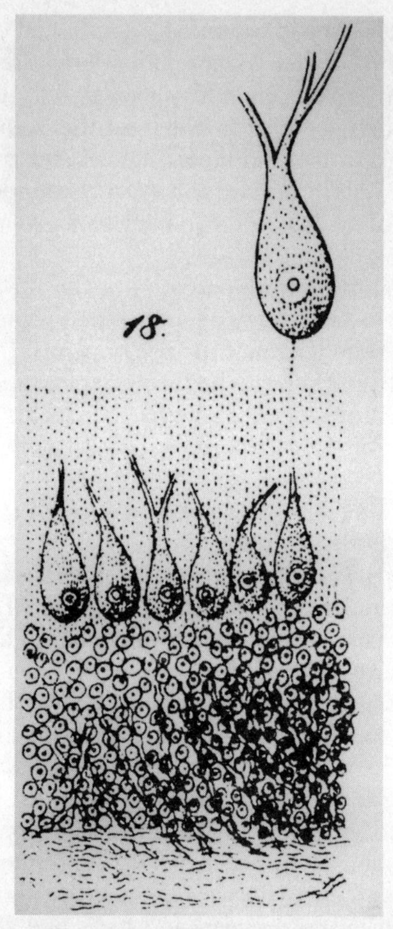

37. Purkyně's illustrations of the large cells in the cerebellum (Purkyně's cells)

relations between the two men. The immediate cause of it remains unknown, but whatever it might have been, it seems probable that part of the problem was that Valentin had difficulty in Breslau in establishing his own scientific persona, a difficulty often encountered by the pupils or collaborators of eminent men. This apparent unease was surely compounded by the fact that he was the son of Jewish immigrants from Poland and thus had no serious academic prospects in Germany. He declined the offer of a chair in Dorpat because a condition of the offer was that he should first undergo baptism. When, in 1836, he accepted the chair of physiology in Bern, where his religious affiliation was not an insurmountable obstacle, he became the first Jew to be appointed to a full professorship (*Ordinariat*) at a German-speaking university.

In the case of Purkyně himself, one can do little more in the compass of one chapter than list his many achievements. The number of eponyms that still bear his name is evidence enough of his productivity. There are Purkyně's vesicle (the *vesicula germinativa* that has already been discussed), Purkyně's cells (the large cells that he discovered in the cerebellum), Purkyně's corpuscles (the lacunae of bone), Purkyně's fibres (anastomosing muscle fibres in subendothelial cardiac tissue), Purkyně's granular layer (branched spaces in the enamel of teeth), Purkyně's figures (dark lines produced by the retinal vessels under certain conditions of illumination), Purkyně's images (three pairs of images seen in the pupil). These eponyms reflect a lifetime of discovery not only in the field of microanatomy but, after he returned to Prague, also in the field of visual physiology. The microanatomical successes of his school in Breslau were not simply due to the availability of better microscopes. Purkyně's technical inventiveness provided methods that made microscopy possible on tissues that had previously been inaccessible, and for those animal tissues that were already accessible he devised procedures for making thin sections that were far superior to those previously available. Mention has been made of his use of prior decalcification for the study of bones and teeth. To enhance microscopic resolution, he also used artificial digestion of tissues and different methods of fixing, staining and mounting the preparations. For recording what he had observed, he employed the recently introduced daguerreotypes; and under his supervision, his assistant Oschatz constructed the first microtome, although Valentin claimed that the idea originally came from himself. Richard Heidenhain, who wrote a biographical memoir of Purkyně,[16] called his institute in Breslau 'the cradle of histology'.

Of the numerous microanatomical observations that Purkyně made, none attracted so much European attention as the discovery, made with Valentin, of ciliary movement in the cells of mammals.[17] In a retrospective memoir that he wrote in 1857,[18] Purkyně recalled that in the spring of 1833, working with frog embryos, he noticed that vibratory cilia were initially present all over the embryo, then, as the animal developed, only over the head, and eventually only over the gill arches. Valentin, together with Adolph Bernhardt, who was then an undergraduate, was at that time studying fertilization of the mammalian ovum. During the course of this work, which formed the subject of Bernhardt's thesis, Valentin noticed that in the ampulla of the oviduct of a squirrel there was a movement of granules close to the mucous membrane. Valentin was unable to interpret this phenomenon, but Purkyně, who was thoroughly familiar with ciliary movement in invertebrates and in the lower vertebrates, at once made the correct diagnosis.

If Purkyně's memory was accurate on this point, it was another case of a discovery made by Valentin and Purkyně together, the former making the observation and the latter interpreting it correctly. Both men continued to work intensively on ciliary movement and established that in higher vertebrates it was to be found in the digestive, respiratory and reproductive systems. At a later stage Purkyně demonstrated that it was also present in the cavities of the central nervous system.[19] In 1835 Purkyně and Valentin demonstrated that the beating of cilia was not dependent on the nervous system[20] and that it continued after the death of the animal. This point met with scepticism on the part of Johannes Müller, but Purkyně held his ground and was proved right in the end. K. E. Rothschuh, an eminent German historian of physiology, takes the view that Purkyně's work, and especially his work on ciliary movement, marks the transition between 'histomorphology' and 'histophysiology';[21] but this apparently ignores the prior contributions of Dutrochet, whose studies on the physiology of the cell, rather than its anatomy, seem to be a much earlier point of transition.

Well before Schwann published his monograph, Purkyně and his school were undoubtedly familiar with the presence of cells in many animal tissues and, on several occasions, they alluded in print to the similarity between animal and plant cells. But the central and specific piece of evidence on which Studnička and his followers base their case in support of Purkyně's priority are lectures that he gave to a meeting of the Society of German Naturalists and Doctors in Prague in September 1837.[22] In his lecture to the first session of this meeting, on 19 September, Purkyně surveyed the tissues in which 'Körnchen' with central nuclei had been found. He first mentions salivary glands, pancreas, wax glands of the ear, kidneys and testes, but then goes on to say that the cells of these glandular tissues show a clear homology with those of the epidermis and those of both mucous and ciliary membranes. Add to these the spleen, thymus, thyroid and lymph glands, which are largely composed of 'Körnchen', then 'The animal organism can be almost entirely reduced to three principal elementary components: fluids, cells and fibres.'* And he goes on to say: 'The basic cellular tissue is again clearly analogous to that of plants which, as is well known, is almost entirely composed of granules or cells.'* (Note that Purkyně here uses 'Körner' and 'Zellen' interchangably.)

Purkyně was intimately familiar with the microstructure of plants, for his first experimental work was done on plant material. He was particularly concerned then with the dynamic character of elastic fibres and their role in causing the anther to burst and thus release its pollen. It is therefore reasonable to say that from the very beginning he was interested not only in structure but also in function. This is a matter of some importance, for, as mentioned earlier, Schwann, in the defence of his own position, dismisses mere appearances that suggest a similarity between plant and animal cells and stresses the functional significance of the model that he and Schleiden proposed. Schwann's criticism can hardly apply to the observations made by Purkyně in his lectures in Prague, first because the whole tenor of what he had to say was concerned with the biological significance of the cells that he saw in both animals and plants; and second, because, in the talk that he gave to the same gathering on 23 September, he specifically discussed the function of nerve cells as well as their architecture. This lecture deals with the cells in the choroid plexus and the ganglion cells that are found in various parts of the brain. Of the ganglion cells he says:

> With regard to the significance of the ganglion cells, it is to be noted that they are probably central structures whose relation to the fibres in the brain and in the nerves can be compared with power generators and power cables. Their entire threefold concentric organization supports this view. Thus ganglia transmit power to the ganglionic nerves, and the cerebral masses transmit power to the spinal cord and the cerebral nerves. They would then be the organs where nerve impulses are collected, generated and distributed.'*

A much reduced plate accompanying these lectures illustrates the stomach glands which are composed of cells that the text compares with plant cells and the 'Purkyně' cells of the nervous system.

It can hardly be doubted that Purkyně's comparison between plant and animal cells refers not only to their appearance, but also to their fundamental biological nature; but he does not, in his Prague lectures, make any mention of how these cells might be generated. Why it should turn out that Purkyně, who argued that all solid animal tissues were composed essentially of cells and fibres, and who referred specifically to the homology between animal and plant cells, was not given appropriate acknowledgement, whereas Schwann was hailed as a pioneer, is an interesting question, to which we now turn.

CHAPTER 10

Müller, Schleiden and Schwann

At the end of November 1832 Carl Asmund Rudolphi, Professor of Anatomy and Physiology at the University of Berlin, died, and proceedings were at once set in motion to fill the vacant chair. The manner in which Johannes Müller engineered his own election to the post throws more than a little light on his character. After a period of intense activity, first as a *Privatdozent* and then as an *Ordinarius* in Bonn, an activity so intense that it required a long intermission to permit his recuperation from an illness that bordered on insanity, Müller was most anxious to go somewhere else: among other things, the intellectual horizons of Bonn were too limited. Nonetheless, he declined the offer of a prestigious chair in Freiburg, explicable only if one assumes that he knew of Rudolphi's terminal illness and had his eye on the Berlin chair. This assumption is supported by the fact that his patience ran out when the chair, declined by Tiedemann and Carus, remained unfilled. Instead of waiting for the invitation (*Ruf*), as was customary, Müller wrote directly to the minister responsible, von Altenstein, and urged his own eminent suitability for the post. This self-advertisement met with some surprise and prompted Du Bois-Reymond to comment: 'wie fern solche Selbstempfehlung auch sonst unserer Sitte liegt' ('how remote from our normal practice such self-advertisement is').

Despite this adverse reaction, Müller's ability and energy were well enough known to gain him the chair, and he gladly took up his appointment in 1833. There were, however, no laboratories for experimental work at the anatomical institute, and the distinguished Berlin school that he founded was for many years obliged to function in cramped rooms within the main university buildings. Although Müller had been a successful academic politician in Bonn, he never did succeed in winning a new institute for himself in Berlin.

Müller's conception of the fine structure of animal tissues was initially rather conventional. As mentioned earlier, he propounded views in early editions of his *Handbuch der Physiologie*[1] that are little different from those advanced by Milne-Edwards and accepted, with modification, by Dutrochet. But in his monograph on the comparative anatomy of the *myxinidae*,[2] which appeared in 1835, he offered a systematic analysis of tissue structure that was based on direct observation rather than speculation. The passage in which Müller describes the similarity between cells in the notochord and plant cells has already been quoted. This observation may be regarded as the starting

38. Johannes Müller (1801–58)

point of the train of ideas that led eventually to Schwann's theory of the cell. Studnička provides evidence that the cells in the notochord were observed earlier by Valentin, but nothing seems to have flowed from Valentin's observation, whereas Müller's observation initiated the stream of comparative histological work for which his school became famous.

39. Jacob Henle (1809–85)

Müller was fortunate in his choice of early disciples. Jacob Henle was a mere fifteen-year-old when Müller first met him on a visit to his home town, Koblenz. Henle, however, disliked Berlin and left in 1840 to take up a chair in Zürich. Henle's position as assistant at the anatomical museum in Berlin was then filled by Theodor Schwann. Both Henle and Schwann retained a lifelong loyalty to Müller, which was reciprocated by Müller's loyalty to them. In an annual report that he wrote for *Progress in Anatomical and Physiological Sciences for the Year 1837*, which appeared in his own Archiv in 1838,[3] Müller made a surprisingly brief reference to Purkyně's 'Körnchentheorie', but he included an account of Henle's as yet unpublished work on the constitution of epithelia and made mention of Schwann's work which was not to appear in print until 1839. Müller did not omit to emphasize the difference between the terminology employed by Purkyně and that of the Berlin group: Purkyně used the rather archaic term 'Enchym' to describe the epithelial layer and noted that it was full of 'Körnchen'; Henle simply described it as multicellular. As soon as Schwann's monograph appeared, Müller devoted eight pages to it in his Archiv and reviewed it in detail in the abstracting journal, *Frorieps Neue Notizen*. In his own monograph on malignant tumours[4] which was published in 1838, and which contains what appears to be the first description of multinucleate cells, Müller notes the presence of 'Zellen' in many tissues, but he makes no mention of Purkyně's view that most, if not all, animal tissues were composed of 'Körnchen', fluids and fibres. It is true that none of the Berlin group was present at Purkyně's Prague lectures, but since Müller's monograph rested on work done in 1837 and 1838, it is difficult to believe that he was unaware of Purkyně's position.

It is central to an understanding of Schwann's contribution that, in defending his own work against the priority claims of others, notably Valentin, he specifically denies the importance of superficial similarities in the appearances of plant and animal cells: 'Such comparisons, however, had no further implication, because they were simply similarities in structures that show the most diverse appearances.'* And he emphasizes that it is the principle of development (Entwicklung) that unifies the concept underlying his whole work. This concept can hardly be discussed without first considering the work of Matthias Schleiden, a view reflected in the fact that most textbooks still refer to the cell theory of Schleiden and Schwann. In a speech given at the jubilee celebrations to mark his appointment to a professorship at Liège, Schwann recounts that he was stimulated, by a conversation he had with Schleiden over dinner, to take him back to his small Berlin laboratory and show the nuclei in the gill cartilage and *chorda dorsalis* of *Rana esculenta* embryos. 'From that moment on,' he claims, 'I devoted all my energies to demonstrating the pre-existence of nuclei in the formation of cells.'*

Schleiden's 'Beiträge zur Phytogenesis'[5] does not make pleasant reading, and it is an indication of what was acceptable in nineteenth-century scientific writing that the article was not only widely quoted, but was even regarded by some, including Schwann, as a masterpiece. It begins with an historical intro-

40. Matthias Schleiden (1804–81)

duction that we would now regard as a travesty. Mention has already been made of Schleiden's disparaging remarks about Raspail and the apparently groundless imputation that Grew manipulated the Royal Society in order to gain priority of publication over Malpighi. Schleiden does praise Robert Brown but adds, without naming them, that others had observed the nucleus before him. He also refers to the work of Meyen on the plant cell nucleus. This was worthy enough, and well known among botanists in Germany, but it could hardly lay claim to priority. It is perhaps the case that Schleiden was the first to draw attention to the importance of the nucleolus in plant cells; but the discovery of this organelle in animal cells was made by Wagner three years earlier.

Since the model proposed by Schleiden and adopted, with hardly any modification, by Schwann, turned out to be immensely persuasive, the key sentences in Schleiden's paper are worth quoting verbatim. He begins his discussion of the cell nucleus by renaming it the 'cytoblast'. This, of course, begs the question, for the word 'cytoblast' at once implies that the nucleus is the structure that generates the cell, and this is precisely what his paper sets out to prove. Then he goes on to say:

> In both places the above mentioned small slime granules very soon arise in the gum and cause the hitherto homogeneous solution of gum to become turbid, or, if these granules are present in larger amounts, to become opaque. Then there appear within this mass separate nucleoli which are larger and more sharply defined. And soon afterwards the cytoblasts emerge showing up as granular coagulations surrounding the nucleoli. – The cytoblasts in this free state continue to grow considerably. . . .*

Cells arise from the nuclei, he argued, in the following way: 'As soon as the cytoblasts have attained their full size, a delicate, transparent vesicle is formed on their surface. This is the young cell which to begin with appears as a very flat segment of a sphere, with its planar side constituted by the cytoblast and its convex side by the young cell which is superimposed on it much like a watch glass on a watch.'* 'Little by little the whole cell now grows out over the edge of the cytoblast and soon becomes so big that the latter eventually appears as no more than a small body enclosed within one of the parietal walls.'* The viscous solution or slime (Schleim) that thus generates the cell Schwann later named the 'Cytoblastem'.

As far as the ultimate fate of the nucleus is concerned, Schleiden advances the following proposition: 'You can still find the cytoblast enclosed within the cell wall, where it participates in the whole of the life cycle of the cell that it has generated. In cells destined to undergo a higher form of development, the cytoblast may either be found in its place, or, after it has been discarded as a useless component, may be dissolved and resorbed within the cavity of the cell.' The nucleus is thus seen as a dispensable structure, once its essential function in the generation of the cell has been discharged.

I think it is fair to say that no part of the scheme proposed by Schleiden turned out to be correct. Baker offers the extenuating circumstance that, in deciding to work mainly with endosperm, Schleiden could hardly have made a worse choice, for the formation of a syncytium in the endosperm and its subsequent division into cells could well have given a totally misleading impression. Nonetheless the fact remains that the model advanced in Schleiden's historic paper is based entirely on misguided speculation, or, at best, on a futile attempt to deduce the function of cell organelles from their structure, in this case far from meticulously observed. In a later work that appeared in 1842,[6] when there was already some criticism of Schwann's monograph, Schleiden took up a less dogmatic position. He questioned the observations of Mohl and Meyen, who had described the generation of new cells by the formation of a central partition within existing cells, but he did concede that he had seen this process in the parenchyma of some cacti. However, because he saw a cytoblast on both sides of the partition, he still held to the view that the new cells were formed in the manner he had previously described.

Of the fifty odd years of Schwann's scientific career, only five, 1834–39, have much relevance to a book dealing with the cell doctrine, for it is only in these years that Schwann occupied himself with the microstructure of animal tissues. He became an assistant to Müller in 1834 and stayed within that ambience until 1839 when he was offered, and accepted, a chair in Louvain. Although the centrepiece of his activity in Berlin was the work that resulted in his famous monograph,[7] he was not, during this period, exclusively engaged in the study of microanatomy. In 1836 he published in Müller's *Archiv* an article on digestion,[8] and in 1837, in *Poggendorffs Annalen*,[9] another on the fermentation of wine and its relevance to the question of

41. Theodor Schwann (1810–82)

42. Title page of Schwann's monograph, *Mikroskopische Untersuchungen*

Mikroskopische

Untersuchungen

über

die Uebereinstimmung in der Struktur und dem Wachsthum

der

Thiere und Pflanzen

von

Dr. *Th. Schwann.*

Mit vier Kupfertafeln.

Berlin 1839.

Verlag der Sander'schen Buchhandlung.

(G. E. Reimer.)

spontaneous generation. The work on digestion led eventually to his discovery of pepsin, and the work on spontaneous generation to a materialistic view of tissue formation. The latter was not only a departure from the vitalistic theories then held by Müller; it also raised serious existential doubts in Schwann's own profoundly Catholic mind. There is no doubt that Schwann was a powerful experimenter. When he moved to Belgium he produced a continuous series of discoveries that would have given him an honourable place in the history of physiology, even if he had done nothing in the field of tissue microstructure. He demonstrated the indispensability of bile by establishing a biliary fistula; he constructed an egg incubator with a primitive thermostat; he devised a bimetal thermometer; he spent twenty years developing two kinds of breathing apparatus with inbuilt lamps; and he invented pumps for the removal of water from mines. It is almost as if the work on the cell theory which is the mainstay of his great fame was a parenthesis in his experimental life, much as, for example, the work on penicillin was a parenthesis in Florey's life.

Schwann's work on spontaneous generation has a special interest in the present context because it bears directly on his model of cell formation. Following the critical experiments that Spallanzani[10] had presented some seventy years earlier in order to refute the ideas of John Needham and Buffon, the question that occupied the minds of believers in spontaneous generation was whether the procedures that Spallanzani had used to exclude external contamination might not have destroyed some life-giving property carried by the air. Spallanzani had heated both the infusion that was to generate life-forms and the air with which it came into contact, and the argument was naturally made that the heat itself might have destroyed some essential vital force. Franz Schulze improved on Spallanzani's experiments by passing the air through sulphuric acid or potassium hydroxide.[11] But still the doubts persisted. Schwann's contribution to the question was a further development of the work of Spallanzani and Schulze. Schwann pre-heated the air before allowing it access to the heated infusion and showed that even when the heated air was passed over the infusion for several weeks no putrefaction or generation of infusoria occured. Moreover Schwann measured the oxygen content of the heated air and found that it was 19.4 per cent, so that the 'vital principle' could not be oxygen as Gay-Lussac had suggested. And he showed that a frog could survive and breathe in this air. Fermentation thus appeared to be due to some micro-organism that was killed by heat. Schwann found that it was not inhibited by *nucis vomicae spirituosum* which killed 'infusoria', but that it was inhibited by potassium arsenate which killed both 'infusoria' and moulds. On examining the fermented material, he noted the presence of 'Körnchen' which resembled, but were not identical to, brewers' yeast – a diagnosis that was confirmed by Meyen.

Schwann had no doubt that spontaneous generation did not occur, but he was very guarded in his conclusions. These appeared not only in *Poggendorffs Annalen* but were also reported verbally by Schwann himself at a gathering of naturalists (Versammlung Naturforscher) in Jena in 1836 and by Müller at a meeting of the Society of the Friends of Nature Study (Gesellschaft naturforschender Freunde) held in Berlin in 1837. It is therefore a matter of astonishment that Schwann should have accepted Schleiden's generative model holus-bolus, for this model was in a very real sense a form of spontaneous generation and was so regarded by Remak, who opposed it from the beginning for this very reason, and by Virchow, who referred to it as the work of the devil. But at some point Schwann must have had serious doubts about what he was advocating, for before publishing his monograph he submitted the manuscript to the Archbishop of Malines (the primate of all Belgium) in order to be reassured that it was not contrary to the doctrine of the Catholic Church. At that time the archbishop raised no objection to the work.

Preliminary accounts of the contents of the famous monograph appeared in three issues of *Frorieps Neue Notizen* in January, February and April of the year 1838;[12] and, no doubt to establish priority, the first two sections of the book were deposited with the Académie des sciences in Paris in the same year. The whole work appeared in 1839.[13] It is divided into three sections to which are added a discussion of the significance of the 'Keimbläschen' (Purkyně's *vesicula germinativa*) and a reply to Valentin's priority claim. The first section deals with the structure and growth of the cells of the notochord and of cartilage. Schwann begins his exposition with these tissues, first because in them the cellular architecture most closely resembles that of plants; and second because, in his view, they present favourable material in which to demonstrate the formation of cells from the 'Cytoblastem', initially via the nucleoli and then via the nuclei. Especially in cartilage Schwann purports to see naked nuclei. The second section bears the dramatic title 'Ueber die Zellen als Grundlage aller Gewebe des thierischen Körpers' (On cells as the foundation of all tissues in the animal body). Purkyně had, of course, seen cells in many tissues and had speculated that they might be the fundamental subunits of most tissues; but Purkyně nowhere makes the categorical assertion that Schwann allows himself in the title of the second section of his book. The rest of the work is very largely a compilation of histological evidence in support of this thesis. Schwann discusses the development of the egg; free cells, among which he includes lymphocytes, erythrocytes, mucous cells and the cells that form pus; cells attached to each other (epithelium, pigmented tissues, nails, claws, feathers and the crystalline lens); cells in which the walls are amalgamated or merge with the intercellular substance (cartilage, bone and teeth); cells that give rise to fibres (connective, tendinous and elastic tissue); and finally, cells with walls and cavities that fuse together (muscles, nerves and capillary vessels). Notably he makes no mention at all of glands, which were the object of special study by Purkyně and his pupils.

The third section of the book is a historical review of theories about the origin of cells and their development. Studnička, having made a special study of Schwann's histological work, concluded that of the twenty-six observations recorded by Schwann, only seven were original.[14] Moreover, as discussed in

Chapter 9, the correspondence between animal and plant cells had been noted on several previous occasions by other investigators. There were, of course, errors in some of Schwann's interpretations of histological appearances. He thought, for example, that muscles, nerves and capillaries arose either by the elongation and cavitation of single cells, or the coalescence of cells; and he regarded fibres as the end-products of cellular decomposition. However, it is clear, from the very beginning of his monograph, that Schwann does not take his stand on the accuracy or novelty of his histological observations. It is the establishment of a general principle underlying all cell development, in plant as well as animal tissues, that he regards as his essential contribution. To quote from his foreword:

> The aim of the present treatise is to establish the intimate connection between the two kingdoms of the organic world by demonstrating the identity of the laws governing the development of the elementary subunits of animals and plants. The main outcome of the investigation is that a common principle underlies the development of all the individual elementary subunits of all organisms, much as the same laws govern the formation of crystals despite their differences in shape.*

These laws were, of course, the ones proposed by Schleiden and, as previously mentioned, they were entirely false. The idea that biological forms might arise from organic matter by a process akin to crystallization was one with which naturalists were familiar. As early as 1810, Georg Prochaska[15] had written: 'When the components of a mixture of animal material changes from the fluid to the solid state, they unite to form small fibres and lamellae. This transformation is produced by the mixture of endogenous centripetal and cohesive forces with those modified by external conditions, so that the process may be regarded as a form of animalistic crystallization.'* Prochaska was very much an eighteenth-century figure who lectured in Latin until he retired. History acknowledges him as an eminent ophthalmologist who, among other things, first accurately described the neurilemma, the striations and sarcolemma of voluntary muscle and the olivary bodies. His general views on the composition and formation of animal tissues were well enough known at this time.

It was all very well for Schwann to enunciate general principles, but when confronted with particular situations he found himself in difficulty. He had argued that one major difference between plants and animals was that cell formation in the former was avascular, whereas in the latter it required the presence of blood vessels. But he had to concede that the generation of cells in the development of the animal egg took place without blood vessels, and he was obliged to refer, albeit briefly, to the work of Henle[16] on multicellular epithelia in which cell formation was thought to be avascular. And when he came to look for his 'Cytoblastem', he found another apparent disparity between the formation of cells in animals and that in plants. He asserts that intracellular (endogenous) cell formation occurs only in animals and quotes Schleiden as never having seen it in plants. Yet in cartilage and epithelium new cells are said to develop in an intercellular 'Cytoblastem' (that is, exogenously). He allows that new cells may sometimes be formed by partitioning of old ones, but he offers no observed example of this phenomenon and does not make much of it. So, whatever Schwann's general principle might be, his monograph actually admits that cells can be generated in a variety of ways.

Schwann's historical review and the annexe in which he defends himself against Valentin's priority claim are, both of them, tendentious and selective. Although he can hardly avoid mentioning the work of Purkyně and Raschkow, whose technique of preparing bone for histological examination had been imported from Breslau to Berlin, he makes no mention of Purkyně's general thesis that, apart from fluids and fibres, all animal tissues are composed of 'Körnchen'. In his annexe, Schwann does indeed quote Valentin's *Handbuch der Entwicklungsgeschichte* of 1835,[17] verbatim and at length, but he does not refer to observations made elsewhere by Valentin and not described in the *Handbuch*. And he dismisses Valentin's comparison of plant and animal cells as an example of fortuitous morphological resemblance, of no consequence because it does not lead to the formulation of a general law. Nonetheless, despite the failings of Schwann's monograph, there was no other place in which all the histological observations had been assembled in a single book, and nowhere else had these observations been united by a single theme, even if that theme proved in the end to be totally erroneous.

Valentin's reaction to Schwann's book was that it contained nothing new. This was obviously incorrect, but understandable in the light of Schwann's cursory attention to most previous work. In support of his own claim to priority, Valentin asserts that he had noted the presence of cells in many tissues, including cartilage and the *chorda dorsalis*, as early as 1835. He does not, as I have mentioned earlier, cite the unpublished Paris prize essay, but he does refer to the article that he wrote in 1839 for Wagner's *Lehrbuch*.[18] Since this was not published until three years later, it is altogether possible that Schwann did not know of it. It bears the title 'Grundzüge der Entwickelung der tierischen Gewebe' (Principles of development of animal tissues) and thus establishes beyond doubt that Valentin was interested in general principles and not simply fortuitous morphological resemblances. It is, of course, true that Valentin did not extract from his own observations the general principle advocated by Schwann, but that, in the light of subsequent events, speaks in Valentin's favour.

In 1840 Purkyně produced for the Silesian Society for National Culture[19] an article entitled: 'Ueber die Analogien in den Struckturelementen des thierischen und pflanzlichen Organismus' (On the analogies between the structural elements of animals and plants). In this article Purkyně says that after the work of Wendt and Kroker on the epidermis of animals and plants, the analogy between the two was obvious, and this was reinforced by the work of Fränkel on the epithelium of the gums. On Valentin's prize essay his precise words were: 'The question posed by the Paris prize on this same topic prompted a thoroughgoing comparison of these structures.'* It is in the article to the Silesian Society that Purkyně elaborates on the different connotations of 'Körnchen' and 'Zellen'. Plant cells, he points out, have a fluid interior and their solid parts are attached to the cell wall, whereas animal cells are solid throughout. He therefore considers it appropriate to refer to 'Zellen' in the case of plants, but insists that 'Körnchen' is a better term for animal cells. He never defected from this point of view.

Purkyně regarded Schwann as a newcomer to the field and referred to him as such. He reviewed Schwann's book in 1840[19] and began his review by pointing out that similar analogies between animal and plant cells had been made by both earlier and contemporary authors. As for himself, he claimed, he had been familiar with this idea as early as 1834 and had imparted it to his pupils who mentioned it in their writings, especially Valentin in his Paris prize essay. Purkyně had two main criticisms to make of Schwann's book. The first was specific. Schwann regarded all structures, including muscles, capillaries and nerve fibres, as modifications of cells. Purkyně, on the other hand, divides the body into the three different categories that have been described, namely fluids, fibres and cells, but he reserves his position with respect to their mode of formation: 'In order to guide theory in the right direction, the mode of formation of muscle and nerve fibres still awaits more empirical data than are at present available.'* The second criticism that Purkyně makes is of a more general kind and is perhaps more telling. He considers that Schwann has been too greatly influenced by Schleiden: 'It appears as if he has been much too carried away by Schleiden's felicitous investigations on the origin of plant tissue, and that he has generalized far too much in taking over analogies from the plant kingdom into the organization of animals.'*

It is of interest that Purkyně refers to Schleiden's work as felicitous ('glücklich') and does not criticize it in so far as it applies to the plant world; it is the application of Schleiden's model to animal tissue that he finds hard to accept. But he does not offer an alternative model of his own. In fact he says very little about the mechanism of animal cell formation, and one has the impression that he is as yet uncommitted. Nonetheless he compliments Schwann on the spirit of his enquiry: 'Schwann's book for the most part leaves us with the impression of genuine scientific enquiry.'* But only 'for the most part' ('grössentheils'), not altogether. Purkyně's review ends with a balanced judgement that could hardly be more diplomatic:

> Moreover, the definite theoretical principles, which make the correspondence between the cell and the nucleated granule so obvious, and which are noted by Schwann himself, retain their validity in a more general sense. Both the idea itself and the merit of having assembled and developed the empirical material to support it in an imaginative and productive fashion remain the unassailable achievement of the author.*

Schwann did not reply to Purkyně's criticism. Having failed to convince minister von Altenstein to create a position for him in Catholic Bonn, he accepted the offer from Louvain and left his cramped laboratory in Berlin. He stayed in Louvain for almost a decade and finally settled in Liège where, in 1852, he moved into No. 11 Quai de l'Université. In this house, which was destroyed during the Second World War, he remained to the end of his life. But, in a long and well-ordered bachelorhood devoted to research, he published nothing more in the field of microanatomy. Some historians have suggested that this silence was due to Schwann's continuing qualms about the possibility that his theory of cell generation might have clashed with Catholic doctrine. I should like to suggest that, despite the receipt of many honours, including the Copley Medal of the Royal Society, Schwann remained silent because he knew he was wrong.

CHAPTER 11

The Reaction to Schwann

Both then and now a theory that eventually proves to be false becomes fashionable and, for a period, prevails against all criticism and circumspection. If we were not so familiar with the phenomenon today, we would be astonished at the reaction to Schwann's monograph. Today we have the unrestrained self-advertisement, the eager courting of the public media and the acclaim of often not very well informed coteries; in the case of Schwann, we have his own skilful advocacy, reinforced, at first, by the enthusiastic support of Müller and his journal, and later by the unquestioning acquiescence of German textbooks. Karl Bogislaus Reichert, a pupil of Müller's, wrote the report in Müller's *Archiv* on progress in microscopical anatomy for the years 1839 and 1840.[1] He does indeed mention Purkyně's 'Körnchentheorie' but mistakes the Körnchen described by Purkyně for cell nuclei, and argues against the generation of cells from 'Körnchen' in the extracellular fluid, a theory that Purkyně at no time advocated. He thus dismisses the 'Körnchen' theory for want of experimental evidence and comes down in favour of the 'Zellen' theory. It appears that he utterly failed to see that Körnchen and Zellen were different names for the same thing. In his book on the development of vertebrates,[2] which appeared in 1840, he is fulsome in his praise of Schwann. Referring to the development of invertebrates he writes: 'But the physiological principle underlying it and our apperception of it became apparent only after Schleiden and Schwann had made their epoch-making discoveries concerning the common mode of life of cells in higher organisms.'* Elsewhere he indulges in some further flattery: 'In his valuable work Schwann has also given us much important information on this subject [the cells in the germinal primordium].'* And in an 1841 paper,[3] again published in Müller's *Archiv*, he refers to Schwann as the creator of the cell theory (Stifter der Zelltheorie), and Purkyně has been elided altogether. The textbooks followed suit. There is barely a mention in them of Purkyně and his school. (Studnička cites Gerber 1840, Bruns 1841, Koelliker 1852, Gerlach 1854 and Leydig 1859, all of which were standard works in their day.)[4]

Although faced with an almost unanimous chorus of praise for Schwann, Purkyně remained sceptical. He was prepared to admit the possibility that Schleiden's model might apply to plants, although he rather doubted it, but he thought it unlikely that it was applicable to the generation of animal cells. In any case, in his judgement, the question required further empirical investigation. He was, of course, well aware of his priority in describing the cellular composition of animal tissues and the correspondence between animal and plant cells, but he recognized that, in the face of the massive support for Schwann, his claim had no chance of succeeding. When he returned to Prague, he turned his attention to other matters.

Henle, who had been Müller's prosector before Schwann's arrival and who couldn't wait to get away from Berlin, presents an interesting case study. He had noticed 'cylinderförmige Körperchen' (cylindrical bodies) in the fluid within the gall-bladder as early as 1835 and reported their presence in an article that he wrote for the *Encyclopaedic Dictionary of Medical Sciences*.[5]* But he did not then appreciate that they were cells that had been shed. It was his systematic study of the epithelia which lined the body cavities that established his formidable reputation. The first account[6] of this work was published in 1837 and a further paper[7] appeared in Müller's *Archiv* in the following year. Henle examined a wide range of epithelial tissues: the respiratory epithelium and its accessories; the conjunctiva; the external auditory meatus and the middle ear; the mucosa of the intestinal canal and the glands opening into it (salivary glands, tonsils, etc.); male and female urogenital mucosa; the glands of the skin; the pleura, the pericardium and the dura mater. It was by far the most systematic and wide-ranging study of epithelia that had so far been made and remained for many years the standard work on the subject.

The epithelia were divided into three categories: 'Pflasterepithelium' (pavement epithelium) that was composed of rather regular flat cells enclosing a nucleus; 'Cylinderepithelium' (columnar epithelium) consisting of a tight array of upright conical bodies with their apexes turned towards the mucosa and their bases towards the free surface. Usually they contained, halfway down their length, a round or oval flat 'Kern mit Nucleus' (This can only mean nucleus and nucleolus, so that apparently Henle used the term 'Nucleus' to describe the nucleolus); Flimmerepithelium' (ciliated epithelium) which was identified only by the presence of the cilia. This study antedates Schwann's monograph and does refer to the observations of Purkyně and Valentin on the choroid plexus. The cells of the epithelium 'contain a round nucleus in which a nucleolus can usually be discerned'.* Again, Henle uses 'Nucleus' to describe the nucleolus, which probably indicates the fluid state of terminology at that time, rather than the confusion in Henle's mind. In any case, his work established beyond doubt the cellular nature of epithelia and indicated that its constituent cells were usually ('meist') nucleated and nucleolated.

In Henle's *Allgemeine Anatomie*[8] we see a dramatic change in attitude. The whole work is permeated by the ideas of Schleiden and Schwann. In the section entitled 'Die Elementarzellen (primäre Zellen, Kernzellen, cellulae nucleatae)' we are given a definitive clarification of terminology:

> In most plant and animal tissues there arise at some point in their development or throughout their whole life microscopic corpuscles of a peculiar and characteristic shape. These are usually denoted by the terms given above. They are vesicles composed of a delicate envelope and fluid contents which are sometimes rather granular. In the wall of the vesicle there is a smaller, darker body, the cell 'Kern', *nucleus*, cytoblast (Schleiden), and this is usually characterized by the presence of one, two and infrequently more, even darker spots that are regularly rounded in shape, the *nucleoli* or 'Kernkörperchen'.*

Then there follows a section entitled 'Die Entstehung der Zellen' (The origin of cells) which begins with a recitation of the views of Schleiden and Schwann. But despite the customary, apparently almost obligatory, obeisance to Schwann, Henle has his reservations. He points out that Schwann's model is based on an analogy with plants and cannot be regarded as established beyond doubt. Schwann argues that the nucleolus comes first, then the nucleus and then the cell. Henle points out that, in his study of the development of the egg, Schwann regards the whole egg as a cell and the germinal vesicle as its nucleus; but Henle considers that there is good evidence to suggest that the vesicle itself is the cell, which leaves its mode of generation uncertain. Moreover, one sees granular material collecting in parts of the cell other than those in the immediate vicinity of the nucleus. In particular, Henle has great difficulty in applying Schwann's model to pus cells which he envisages as formed *in situ* by the transformation of fixed tissue cells; and he regards the lobulated nucleus of the polymorphonuclear leucocyte as a concatenation of several nuclei.

Henle accepts Schwann's view that cells are generated from nuclei, but does not accept that nuclei are necessarily generated from nucleoli. The multiplication of cells in the skin, which apparently takes place in the absence of blood vessels, is adduced as the best example of cells formed *de novo* from nuclei, but Henle doubts whether this process necessarily begins with the nucleolus: 'Other observations make it doubtful whether the granular material which generates the cell nucleus is precipitated only in the vicinity of the nucleolus';* and he gives examples. However, he goes on to say that structures that resemble cells ('zellenähnliche Gebilde') but lack a nucleus, are not convincing evidence against a nuclear origin for cells. 'To begin with, just as there are nuclei without nucleoli, so there are cells without nuclei.'* Overall, Henle's view is that the scheme proposed by Schleiden and Schwann, if one excludes the necessary formation of nuclei from nucleoli, is only one way in which cells are generated. He agrees that where this process does occur, it is akin to crystallization. On the other hand, 'It can no longer be doubted that in the animal body too cells can be formed within cells.'* Henle proposes three mechanisms other than that envisaged by Schwann: 'Sprossenbildung' (germination), 'innere Zeugung' (internal generation) and 'Theilung' (cell division). One has the impression that these are what he saw

43. Franz Unger (1800–70)

or thought he saw, but it is really only in the skin, where appearances are notably deceptive, that he found strong evidence for the generation of cells from nuclei by the mechanisms that Schwann described. Henle also disagreed with Schwann about certain forms of cellular differentiation. In general, he was much more inclined to think that one cell type could be transformed into another. He did not object to the idea that the shape of cells was determined by their position: if they were aligned along a flat surface, he argued, they formed lamellae, but otherwise they formed conical corpuscles. But he did not believe that fibres were necessarily modifications of cells or nuclei: they might be secondary deposits that were formed in the intercellular substance. Furthermore, he thought it unlikely that disappearance of the nucleus was the normal course of events; in most cases the nucleus persisted and underwent a series of changes that he described in minute detail. Henle disagreed with much of what Schwann wrote, but nowhere did he bring himself to say that Schwann's model was basically incorrect. All he was prepared to assert was that there were alternatives. Henle was not a restless man. When he was finally appointed to a chair in Göttigen he remained there for thirty years, and, in the process, became the undisputed doyen of German anatomists. One cannot help wondering whether there were perhaps other reasons for his eagerness, as a young man, to leave Berlin, quite apart from his well-known opposition to Prussian autocracy and his dislike of the city itself.

The first frontal assault on Schleiden's model *in toto*, and not merely its application to animal tissues, came from Franz Unger. A short biography of Unger by Alex Reyer appeared in 1871,[9] but there is otherwise little mention of him in the works of later historians except perhaps in Austria. This neglect is undeserved, for Unger was responsible for more than one fundamental discovery that changed, or ought to have changed, received opinion about the nature of the cell.

Unger was born in the Steiermark and pursued his academic career first in Graz and then in Vienna, where he was eventually appointed to the professorship of plant anatomy and physiology. Since he was sixty-six years old when the battle of Königgrätz (1866) finally decided the political rivalry between Prussia and Austria, he would have regarded himself as a citizen of the proud Hapsburg empire, and he would have had no special respect for what came out of Berlin or, for that matter, Jena. He was certainly a highly original and stimulating teacher who, at one time, numbered Gregor Mendel among his pupils.

The first of Unger's seminal observations emanated from Kitzbühl, where he was engaged in a collaboration with Professor Ettinghausen, a physicist in Vienna, who had access to an excellent Plössl microscope. He noticed that in a squash preparation of the pollen of *Malva sylvestris*, the cytoplasm of the cells showed movements that were not those described by Robert Brown and that we now call Brownian motion. This observation, which was published in *Flora* in 1832,[10] is best described in the author's own words:

> The movements were not oscillatory but at different times advanced, retreated, moved sideways or gyrated. Individual granules either avoided each other or approached each other, and in the latter case their movements became more vigorous. Those that were submerged rose to the surface and those at the surface went to the bottom. This marvellous, or at least astonishing, scene resembled an army of monads full of inner vitality, full of an inner self-determination that revealed itself in their movements.*

Unger took great pains to show that these movements were not the particulate motion described by Brown. Momentary desiccation of the cell or a mere trace of alcohol eliminated them completely. And a very fine powder of glass particles or of other inorganic substances behaved quite differently under the same conditions. Unger was well aware of the importance of his discovery, but it is a pleasure to read the modest and courteous account that he gave of it:

> These few observations and experiments might be enough for the time being to indicate that the microscopic molecular world is still by no means adequately understood, despite the investigations that have yielded so many splendid results, especially in recent times. There is therefore nothing that could be more urgently desired than that Prof. Ettinghausen, who has such a superb instrument in his possession, might find the time to pursue this matter with the attention it deserves.*

44. Ferdinand Cohn (1828–98)

Unger probably did not come to a full realization of the significance of his discovery until Ferdinand Cohn published his definitive paper on this subject in 1850.[11] Cohn is another man who has not been adequately treated by subsequent historians of the cell doctrine. He was born into a Jewish family in Breslau in 1828, and despite his having spent virtually the whole of his academic life at the university there, his promotion had been fraught with difficulties and had come very slowly. He did not acquire laboratory space until he had been a professor for some twenty years. When he was finally allocated some modest space, he turned it into the first institute of plant physiology in Germany, and he himself became one of the foremost European authorities on the classification of bacteria which he regarded as primitive forms of plant life. It was to him that Robert Koch turned for an assessment of the latter's early experiments on anthrax, and it was in the journal that Cohn founded, the *Beiträge zur Biologie der Pflanzen* (Contributions to the biology of plants), as part of a series of papers entitled 'Untersuchungen über Bacterien' (Investigations on bacteria), that Koch first published his work.

In his 1850 paper, which deals with the development of *Protococcus pluvialis*, Cohn makes the point that the consistency of the cytoplasm (Endochrom) is variable, and he is not sure whether this organism has a nucleus; but he considers that contracility is an essential characteristic of the cell cytoplasm, and that this is true for higher plants as well as primitive forms: 'What characterizes the primordial cell most clearly is the contractile element in the organism; and this is significant for the life of plants in general, although it appears most clearly as an essential feature of swarm cells. This contractile element endows the plant cell with the capacity to change its shape without a corresponding change in the cell volume.'*

The idea that a contractile substance ('contractile Substanz') was common to the cytoplasm of both primitive and higher forms was resurrected by Alexander Ecker,[12] a professor of anatomy at Basel, whom Cohn quotes. This substance was said to be most fully developed in muscle and to be present in a primitive form in infusoria, rhizopods and hydroids. Remak mentions that Ecker also claimed to have seen movement in the blastomeres (cells formed by cleavage of the fertilized egg) of the frog, but gives no reference. It was Ecker who pointed out that his 'contractile Substanz' was essentially the 'sarcode' of Dujardin, whose work, it seemed to him, had been largely overlooked. Cohn, like Ecker, refers to Dujardin's 'sarcode' and considers it equivalent to his own 'contractile element'. In his *Die Anatomie und Physiologie der Pflanzen*,[13] which appeared in 1855 and which embodied the results of almost thirty years of experimental work, Unger fully supports the ideas of Cohn and recognizes that it is the contractility of Dujardin's 'sarcode' that is responsible for the cytoplasmic movements that he and Ettinghausen had discovered more than twenty years previously.

Courtesy pervades Unger's writing even when the position he adopts is in direct opposition to that of Schleiden, whose arrogant style offers a striking contrast. Unger first expressed his doubts about Schleiden's model as early as 1841[14] and considered even then that the commonest form of cell generation was that driven by cell division. He did not, however, at this stage, think other mechanisms implausible. But in 1844 he produced a series of papers[15] that advanced arguments in open conflict with Schleiden's views and in favour of the idea that new cells were formed by the division of one cell into two by means of a central partition. In the first of these papers he establishes that growth of the internode is due to the accretion of cells and that they are arranged in rows. In the second, however, after presenting further measurements on the increase in cell size and number in the meristematic tip of *Campelia Zanonia*, he comes to consider the general question of how cells are formed:

> The second question, namely in what way the generation of new cells takes place in a tissue that has already been formed, is necessarily linked to the view one has about cell formation in general.*
>
> I think that to some extent I never accepted the view that the cytoblasts were the source of new cells in this sense that the latter were the direct spatial product of the former; and especially in the case that I have discussed, it would be very difficult to explain the formation of new cells in such internodes, for their cells usually have no nucleus. But my main argument against this theory is that one cannot observe young cellular vesicles emanating from the cell nucleus, at least not where new growth occurs.*

Unger is thus aware that the apparent absence of the nucleus is not itself a decisive argument against Schleiden's scheme and makes it clear that his main objection to it is the lack of empirical evidence in its support. He then goes

45. Carl Nägeli (1817–91)

on to describe the mechanism of cell formation that he can actually observe. He notes that the thickness of the cell wall is variable and that some cells have very thin walls. 'If one examines the formation of these delicate walls more closely, one cannot fail to see that they usually act as partition walls for cells expanding in one direction or another, so that they are, so to speak, divided into two compartments.'*

This is, of course, the mechanism first described by Dumortier in *Conferva*, but Unger does not mention him, perhaps, in the light of his habitual courtesy, out of ignorance. But he does mention the previous work of Mohl and Nägeli who described cell division in *Conferva* and *Marchantia*, but not, and this is crucial, in higher plants. Unger's views on the nature of the 'Scheidewand' (partition wall) met with some criticism from Nägeli. Unger believed, as a result of his studies on the hairs of the young leaves of *Syringa vulgaris*, that the partition wall was, at least initially, single, whereas Nägeli insisted that it was composed of two lamellae. Moreover, Nägeli defends the thesis advanced by Schleiden that the nucleus may disappear when its essential function has been discharged. Unger replied to Nägeli's criticisms in the last paper of the series, but considered, rightly, that whether the partition wall was single or double was a far less important question than the mechanism by which new cells were formed. His final answer to Nägeli was that 'In most cases where there is growth in the mass of the cellular tissue, this takes place not through intrautricular, but through meristematic cell formation, so there can be no question of there being mother cells or of their dissolution.'* Unger nowhere categorically denies the possibility that other forms of cell multiplication might, under certain circumstances, exist, but he has not observed them; and he is quite clear that in higher plants, just as in more primitive forms, binary fission, achieved by the formation of a partition wall, is the normal mechanism involved.

It would be surprising to a twentieth-century eye that Nägeli's papers on cell formation were much more influential than Unger's, were it not for the fact that the message transmitted by Nägeli could more easily be accommodated within the theory then fashionable than Unger's views, which challenged that theory head on. For although Unger accepted the possibility that modes of cell multiplication other than cell division might, in principle, exist, it is clear from his papers that he was fully committed to the view that cells were not generated *de novo* out of an undifferentiated organic 'Cytoblastem' but were formed by the division of existing cells into two. Nägeli, on the other hand, while admitting the existence of binary fission, and indeed stressing its importance in certain situations, presented it as one of several mechanisms which he claims actually to have observed and to which he attaches no less importance. He expounds his views in two key papers, one that appeared in 1844[16] and the other in 1846.[17]

The 1844 paper is mainly concerned with silkweeds and other primitive plants. In these Nägeli observes binary fission and confirms that in *Conferva glomerata* and some other species this occurs only in the terminal cell of the filament, except where branching occurs. But there is no mention of Dumortier. Nägeli is much exercised by the precise nature of the partition wall. Of the two mechanisms then being proposed, 'Abschnürung', in which the cell membrane divides the cell much like an elastic band, and the formation of a 'Scheidewand' (partition wall), Nägeli favours the latter, but does not draw a distinction between the cell membrane and the plant cell wall. As mentioned above, Unger thought this a secondary matter. Nägeli believed that the nucleus simply divided into two and did so before the formation of the partition wall (more will be said about this later), but he could not see nuclei at all in the confervae and entertained the possibility that they might be anucleate. In *Sphacelaria scoparia* (Fucales) he observed an accumulation of granules segregated to one side of the cell and concluded from this that Schleiden's model of cell formation in association with the nucleus could not hold in this species.

Schleiden was, as could be expected, highly critical of Nägeli's views, but Nägeli defended himself vigorously and thought that he had adequately met Schleiden's criticisms: 'I conclude that in the confervae the multiplication of cells cannot possibly occur by the formation of small free cells within the interior of others.'* But he still believed that, in the formation of spores and pollen mother cells, the primary nucleus might generate a secondary one and then be resorbed. Nägeli's position at this stage is clearly revealed in the summary that ends the 1844 paper. He believes that binary fission is the only form of cell multiplication in the confervae and some other rudimentary aquatic plants, except in the case of their germ cells and spores, where special mechanisms of a different kind exist. However in a wide range of higher plants, including phanerogams, he asserts: '. . . that in these divisions of the plant kingdom, with the exception of specialized mother cells, there is found only free cell formation in association with the nucleus'.* At this stage, Nägeli thus sits on the fence. Although he canvasses binary fission by means of a partition wall as the general mechanism of cell formation in his collection of primitive plants, he leaves ample room for Schleiden's model elsewhere.

By the time Nägeli wrote his 1846 paper, he had greatly extended the range of his studies and had further defined his position. He adhered to his view that, with the exception of the germ cells and spores, cell multiplication in the silkweeds and other primitive plants occurred by means of binary fission. He again rejects Schleiden's criticisms. Schleiden had originally questioned not only Nägeli's observations, but, as I have discussed in an earlier chapter, also those of Mohl and Meyen on the significance of binary fission in the cells of confervae. Schleiden had argued that, despite the division of the cell into two, the new cells were formed as he had described. Now he modified his position. Indeed, he kept modifying it in such a way that it eventually became almost incoherent; but for cell multiplication in the confervae, at least, Nägeli continued to rebut his modifications. Nägeli still did not reject *de novo* free cell formation in higher plants, or, indeed, in the specialized cells of confervae. He claims actually to have seen it, and he gives a systematic classification of its manifestations. Two main categories of 'freie Zellbildung' (free cell formation) are given: cell formation in the absence of a visible nucleus and cell formation in the presence of a parietal nucleus. As examples of the former category he lists many algae, including confervae, where 'occasionally a germ cell is formed within a cell entirely out of its own contents'.* Nägeli regarded this, however, as an abnormal form of cell reproduction in these primitive forms. Of the second category he has this to say: 'free cell formation in the presence of a visible nucleus I have so far observed unequivocally in the embryo sac of phanerogams.'* In a special section entitled 'Free cell formation as a general law', he says that this type of reproduction is to be observed most clearly in the germ cells of *Zygnema*, the sporangial cells of *Achlya* and the large cells formed by abnormal multiplication in *Bryopsis*, *Conferva* and other algae. The precise mechanism that he claims to have observed is an unalloyed echo of one part of Schleiden's original thesis: 'Within the contents of the mother cell there arises a nucleus. This attracts to its surface a variable proportion of the contents of the mother cell, which, at least to begin with, consist of homogeneous slime. The whole surface of this part of the cell contents is then invested in a membrane.'* For Nägeli the essential element in all these modes of cell formation is the segregation of the cell contents into two separate entities, a concept that he calls 'Der Begriff der Individualisirung des Zelleninhalts' (the concept of the individualization of the cell contents). He regards as real almost every imaginable mechanism by which this individualization of cell contents might take place, and thus offers all things to all men. It is true that he greatly extended the range of observations on binary fission and stressed its importance, but, in my view, the essential reason for his widespread influence was that he merely rocked the boat, but did not sink it.

Mention must be made of Nägeli's correspondence with Gregor Mendel. Of the many biologists to whom Mendel sent reprints, produced at his own expense, of his paper on the consequences of hybridization in *Pisum*, Nägeli was the only major figure to take the work seriously enough to repeat it. But he chose *Hieracium* with which to do it. This, however, is a plant that reproduces asexually and its progeny did not therefore show the ratios Mendel had found in *Pisum*. When Nägeli wrote to Mendel about this, Mendel himself carried out a long series of experiments with *Hieracium* and was able to confirm that this plant did not behave like *Pisum*. Dunn, in a commemorative essay on Mendel in 1951[18] describes the choice of *Hieracium* as 'unfortunate', but Orel,[19] in a recent biography of Mendel, justifies the *Hieracium* experiments by pointing out that they were merely part of a systematic research programme that was designed to explore the consequences of hybridization in a variety of plants. This may well have been so, but the results obtained with *Hieracium* were apparently enough to make Mendel himself doubt whether the laws of inheritance that he had revealed in *Pisum* were universally valid.

All the work I have so far described in this chapter deals with plant cells, where microscopy permitted a mass of speculations, but also some facts. In the case of animal cells, empirical evidence was fragmentary. The criticisms of other botanists demolished Schleiden's model for plants within a few years; but Schwann, for want of decisive contrary data in animal cells, tried to hang on to his views. Even when it was clear that the model that he had endorsed and propagated did not apply to animal cells either, Schwann attempted to justify his position by arguing that more recent observations nonetheless confirmed the central role of the nucleus in the generation of the cell. The principal contributor to the evidence against Schwann was Robert Remak, whose first observations on the multiplication of chick embryonic red cells were published in 1841.[20] This was a mere two years after the publication of Schwann's monograph and three years before Unger published his series of papers on cell division in plants. Remak noted that new cells were not generated by any kind of endogeny or by *de novo* formation of 'Cytoblastem' in the extracellular fluid; and he failed to see any naked nuclei. The red cells multiplied only by binary fission. This was not a chance observation of Remak's, but a reproducible phenomenon which, in his capacity as *Privatdozent* at the University of Berlin, he demonstrated annually to his students. By 1845,[21] he had seen new muscle cells formed by the division of pre-existing ones in primitive muscle bundles. And by 1852[22] he had come to the view that the only form of cell multiplication to be found in the animal body was multiplication by means of binary fission. Needless to say, the case that Remak made was not at first accepted, and even Virchow, as late as 1854, had his reservations. But Remak won through in the end, although, in his own lifetime, he was never given the credit that such a momentous discovery deserved. Nor has he been given it to the present day.

CHAPTER 12

From Fertilized Egg to Embryo

Although observations on the formation of the embryo within a fertilized hen's egg antedate even the *De generatione animalium* of Aristotle, no connection was made between the early stages of embryonic development and the cells which, after Purkyně and Schwann, were generally accepted as the fundamental subunits of both animals and plants. Nor had it been demonstrated, before the work of de Graaf, that the embryos of mammals also developed from fertilized eggs by a process that was analogous to that which had been observed in the eggs of birds. In 1672, Regnerus (Reinier) de Graaf, a practising physician in Delft, claimed to have seen the mammalian ovum.[1] Here is the relevant passage from his book: 'We therefore examined again and again the passages through which the eggs must pass, and we found in the middle of the right oviduct one egg, and in the tip of the uterine horn on that side two minute eggs were revealed.'* These eggs were illustrated in an accompanying plate, and they were said to have been found on the third day after coitus.

It is the timing of the discovery that aroused the suspicion of Ludwig Wilhelm Bischoff 170 years later. In an essay[2] that won the 1840 prize of the Royal Prussian Academy of Sciences, he refers to this passage of de Graaf's and says that the latter is quite right in describing the ovum as constituted of two adherent vesicles, but points out that his own study of the development of the rabbit's egg makes it clear that this change in the egg never occurs in the Fallopian tubes nor, initially, even in the uterus. Bischoff therefore considered that de Graaf's observation was an artefact. Baker, more than a century later again, was more charitable. He assumed that what de Graaf saw was either a morula or a blastocyst (later stages of development). So there is doubt about whether de Graaf actually discovered the mammalian egg, although there is none about the follicle that bears his name.

There is no dispute, however, about the description of the mammalian ovum given by Karl Ernst von Baer, who reported his observations in a Latin letter written from Leipzig in 1827.[3] For obvious reasons, there was then little hope of studying the sequence of development in a single mammalian egg, still less in a single human egg. Towards the end of the eighteenth century, attention therefore turned to the examination of free-living ova, or at least those that were easily accessible. In 1775 Roffredi[4] described cleavage in the

46. Karl von Baer (1792–1876)

egg of the free-living nematode, *Rhabditis*, but, of course, he did not then associate what he saw with the formation of blastomeres. Spallanzani, in 1780,[5] saw what appears to be the four-cell stage in the development of the ovum in the toad, *Bufo*, and perhaps also the two-cell stage in the green frog. The protuberances were separated by what he called 'solchetti' (small furrows), but Baker considers that what he saw was the neural groove. Spallanzani clearly had no conception of the biological significance of the 'solchetti', for he continued to believe in the pre-existence of a formed micro-embryo within the unfertilized egg.

With Prévost and Dumas (1824)[6] we have the first accurate and seriously analytical description of segmentation in the egg. This will be discussed in some detail, for the manner in which von Baer dismissed this work and the work of Mauro Rusconi seems to me to besmirch von Baer's magisterial reputation. The relevant observations of Prévost and Dumas were made on the developing eggs of the frog which were first treated with fluid expressed from the frog's testicles. After this treatment the egg forms a line which is the first stage in the formation of a criss-cross of furrows:

> This line which, at first, can only be made out on the surface of the egg as a very slight hollow, deepens with unimaginable speed, and sets in train the formation of a substantial number of small parallel wrinkles that run perpendicularly to the line itself and that arise in the furrow which it produces. The latter keeps deepening, and the egg is soon divided into two quite distinct segments. A new line then makes its appearance, but this runs more or less at the boundary that separates the two hemispheres, one brown, one yellow, and cuts the egg in a circular fashion rather like a kind of equator. The brown hemisphere was divided into four equal portions each of which

47. Mauro Rusconi (1776–1849). A self-portrait in pencil

> is cut into two by means of new hollows that run parallel to the furrow that had preceded them. This hemisphere is thus divided into sixteen equal parts or thereabouts. Then the brown part of the egg is divided into a number of granulations rather like those of a raspberry. At first, you can count thirty or forty, but within two hours they have in turn been subdivided and there are more than eighty of them.*

These quotations show first that Prévost and Dumas were not limited in their observations to surface phenomena. They were fully aware that the furrows that they observed on the surface of the developing egg led to its complete division, and that continued repetition of this process resulted in further subdivision until the whole structure resembled a raspberry. If this were not so, it would be difficult to explain their use of the words 'segment' and 'hemisphere' or to imagine what notions they might have entertained about the structure of raspberries. They clearly thought that the surface furrow and the ever deepening hollow with which it was associated cut each progressive subdivision of the egg into two, an idea which survived for more than thirty years after it was first promulgated and which, in the German literature, was given the name 'Abschnürung' (ligation). It goes without saying that Prévost and Dumas had no idea that they were dealing with the multiplication of cells, as Purkyně and Schwann later envisaged them. In 1824 that was hardly possible.

In a book published in 1826[7] which dealt with the development of the common frog, Mauro Rusconi, like Prévost and Dumas before him, accurately described segmentation of the egg. Rusconi was attached in an adjunct capacity to the University of Pavia, but was never appointed to a chair there. In middle age, he retired to private life. His book was written in French and contains the following key passage:

> If one confers a certain degree of firmness on the fertilized egg, either by boiling it or by some other means, and if one carries out this operation at a time when its surface, especially the brown surface, is entirely covered with a criss-cross of furrows, one can then separate the fertilized egg into several masses which vary in size according to whether there has been more or less multiplication of the furrows on the surface; in short, we find, by repeating this experiment at different times, that the whole substance of the fertilized egg first divides into two, then into four, and each part goes on dividing and subdividing into ever smaller units.*

Eventually, the whole structure becomes 'une masse granuleuse' (a mass of granules). There is thus no doubt whatever that Rusconi was fully aware of what was going on; and the passage quoted is a perfectly accurate description of segmentation in the fertilized egg. It is therefore a matter of more than a little surprise that von Baer in his notable paper on metamorphosis in the eggs of anurans[8] summarily dismisses the earlier work.

This paper appeared in 1834, the year in which von Baer moved from a chair in Königsberg to St Petersburg where he served for almost three decades as professor of comparative anatomy and physiology at the Medico-chirurgical Academy and the Academy of Sciences. Referring to the work of Prévost and Dumas, he asserts that it is limited to the observation of the furrows on the surface of the egg: 'their observations on the conspicuous furrows that can be seen on the surface of frogs' eggs'.* And the reason for this limitation to their work was: '. . . doubtless because they were not aware of any method for removing the white of the egg in order to harden the yolk and thus analyse it'.* Von Baer then goes on to say that, since he has the relevant methods at his disposal: 'We therefore propose in all cases to begin with what can be observed on the surface and then proceed to the interior. But so that there may be no misunderstanding, we will observe at the very outset that the slits that one sees on the surface are simply the boundaries of divisions that the whole yolk undergoes.'*

There is no doubt that von Baer made a more thorough analysis of segmentation than his predecessors. He followed the process in detail from the first to the tenth division in the eggs of the brown frog; and the plate accompanying his paper amply illustrates the precision of his observations. But to assert that Prévost and Dumas dealt only with surface phenomena is, if one is charitable, a misreading of their paper, or, if one is less so, a deliberate misrepresentation of it. In any case, von Baer's criticisms can hardly apply to Rusconi, whose work he describes as vague and uninterpretable. Rusconi replied to von Baer in a letter that was published in 1836 in the *Archiv für Anatomie und Physiologie*.[9] This had been written in response to a request from E. H. Weber for Rusconi's views on the criticisms that von Baer had made. In his letter Rusconi quotes the passage given earlier. (Actually, the letter contains two minor changes to the text given in his book, but they do not materially affect the sense.) That passage disposes of the argument that

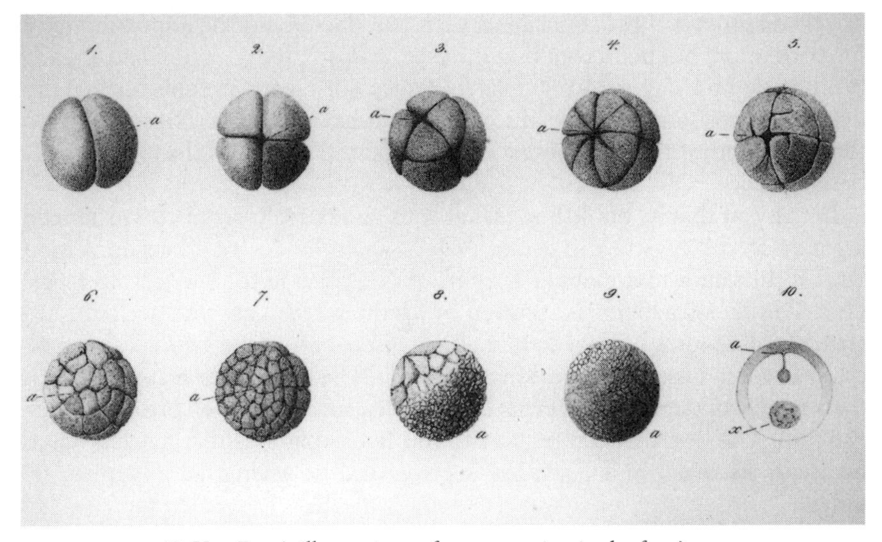

48. Von Baer's illustrations of segmentation in the frog's egg

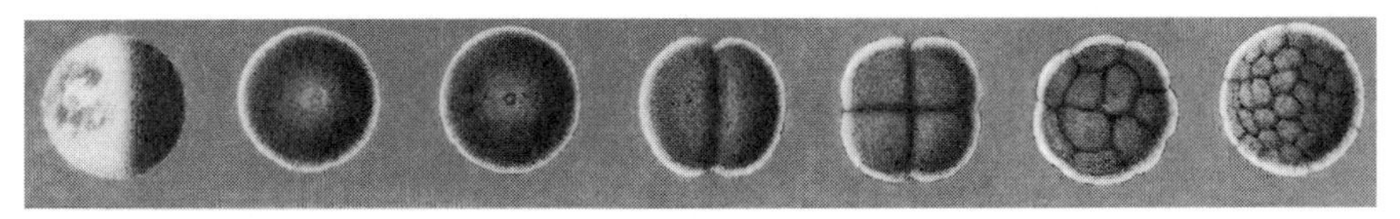

49. Rusconi's illustrations of segmentation in the frog's egg. The original is in colour

the inability to remove the egg-white and harden the yolk had limited previous observations to the surface of the egg. Rusconi had removed the egg-white and had hardened the yolk in more than one way. Moreover, it is abundantly clear that Rusconi had observed segmentation of the egg into subunits of ever diminishing size. And the beautiful hand-painted plates that form part of the book illustrate the process no less clearly than the figures accompanying von Baer's paper.

Since von Baer refers to Rusconi's book, one must assume that he has read it. If this is so, it is difficult to understand how he could have permitted himself the luxury of dismissing it in such a cavalier fashion. But Rusconi gave as good as he got. He expresses astonishment that a scholar as distinguished as von Baer could hold the view that the segmentation of the egg is designed to give all parts of it direct access to the action of the sperm. He points out that by the time segmentation occurs the egg is fertilized and the action of the sperm long over. Von Baer emphasizes the fact that his experiments eliminate the notion of 'preformation' of the embryo (its existence as a fully formed miniature in the unfertilized egg): 'Of all the eggs with which I am acquainted the eggs of the frog seem to me to be the only ones in which "preformation" can be disproved.'* Rusconi agrees with von Baer about the impossibility of 'preformation' but points out that the case against 'preformation' had already been made by Prévost and Dumas. He does not regard it as his business to defend Prévost and Dumas, but he makes it clear that von Baer's assertion that these authors made observations only on the surface of the egg is totally baseless.

In the year that he published his reply to von Baer, Rusconi also produced a paper on the eggs of the perch.[10] Once again he described segmentation after fertilization in a manner altogether comparable to that seen in frogs' eggs. Von Baer's paper was received by German scientists with acclaim, and virtually all textbooks refer to him as the discoverer of the biological significance of furrow formation and segmentation. The true position appears to be that the discoverers were Prévost and Dumas, and, with even greater clarity, Rusconi. The best that can be said of von Baer's paper is that he elaborated on their discovery and provided a great deal of additional experimental detail.

At about the same time as von Baer published his paper on segmentation in the frog's egg, de Quatrefages published the memoir discussed previously on the embryogenesis of freshwater snails and other gastropod molluscs.[11] Whereas von Baer, and Rusconi before him, clearly interpreted what they saw as a progressive partitioning of the egg, Quatrefages interpreted the whole process as an example of endogenous formation of globules within mother cells. It is apparent that Quatrefages went as far with this material as simple observation without manipulation could take him, but it is perhaps surprising that he does not compare his own observations with those of Prévost and Dumas or Rusconi, both written in his own language. His mind is obviously dominated by the ideas on cell generation that were then fashionable in Paris. Von Siebold, in an article written for Burdach's physiology text and published in 1835,[12] makes the comment that the formation of furrows has been observed only in the eggs of vertebrates; it has not, he points out, been seen in invertebrate material. He refers to the work of Prévost and Dumas, Rusconi and von Baer, but he is apparently unaware of the paper by Quatrefages. Von Siebold reported that furrows were to be seen in the eggs of a very wide range of invertebrates including several species of *Ascaris*, *Filaria* and *Strongylus*.

Guided by the analyses of previous authors, von Siebold also interpreted furrowing as a progressive subdivision of the egg. But he remained unsure of the exact significance of the vesicles thus generated. He noticed that within each vesicle there was a smaller one and suggested, in an analogy with what one finds in other animals, that the larger vesicles corresponded to Purkyně's *vesiculae germinativae*, and the smaller ones to the germinative spots of Wagner (nucleoli). Two years previously, in October 1833, Jean Victor Coste reported to the Académie des sciences in Paris that, after an intensive search, he had found on the surface of the yolk in an unfertilized rabbit's egg a vesicle that he believed to be the *vesicula germinativa* that Purkyně had described in the hen's egg. His experiments were summarized in *Froriep's Neue Notizen* in November of the same year.[13] So it appeared that Purkyně's vesicle was present in the eggs of mammals as well as birds, and something like it was also present in the eggs of invertebrates. But there was at that time little understanding of its function.

It is of interest to consider now the influence that Schwann's views had on studies of development in the early embryo. This is perhaps most clearly illustrated in the writings of Martin Barry (1802–55). Barry obtained his MD degree from the University of Edinburgh, but, having a private income, he never engaged in medical practice or held a permanent appointment. In 1837 he was drawn to Germany by the facilities offered by Müller, Ehrenberg and Wagner. He actually worked with Schwann and later corresponded with him. Barry published a series of papers entitled collectively 'Researches in Embryology' in the *Philosophical Transactions of the Royal Society* between 1838 and 1841 – a period that straddles the publication of Schwann's monograph. In the 1838 paper,[14] he discusses the vesicle of Purkyně, which he has examined in the dog, cat, ox, sheep, rabbit, common fowl and pigeon. He attacks von Baer's view of the function of this vesicle, but his own observations are purely descriptive. There is as yet no talk of the division of the nucleus or nucleolus. By 1839,[15] however, he suggests an idea that is a clear derivative of Schwann's views: 'The stages of the mammiferous ovum (illustrated by a plate), presenting as they do successive sets of vesicles, each set consisting of smaller and more numerous vesicles than the last, suggest the idea that in the interior of each vesicle there arise two or more infant vesicles, the parent vesicle in each instance disappearing by liquefaction.' He then goes on to propose a model which is pure Schwann and which involves the nucleolus. And in a postscript he says: 'Such really is the case, as proved by the condition of the ova which I have met with since.'

In the paper published in 1840,[16] Barry concludes that Purkyně's vesicle is a cell nucleus; the section dealing with this question has a subheading entitled 'The Rudimentary Embryo is the Nucleus of a Cell'. His exposition is still largely based on the views of Schleiden and Schwann, but some criticism of these views has begun to creep in: 'The existing view, namely that a nucleus, when it leaves the membrane of its cell, simply disappears by liquefaction, is inapplicable to any nucleus observed in the course of these investigations.' At this stage Barry considered that it was the centre of the nucleus that was critical for further development. There are three papers in 1841,[17] one of which deals with the *chorda dorsalis*[18] (the classical material of Müller and Schwann) and another with the corpuscles of the blood.[19] Barry emphasizes again the central role of cell nucleus in the development of the embryo: 'The origin of the embryo from the nucleus of a cell may assist to solve a question on which, I believe, physiologists are not agreed.' Red cells, however, are, in his view, propagated by division of the cell nucleus. According to Baker, who may be showing an excess of enthusiasm in this respect, Barry was the first to equate blastomeres with real cells. Certainly, he did mention that 'Some of the vesicles presented in their interior a minute pellucid space which might possibly have been a nucleus.' But his painstaking studies are marred by an unconcealed effort to fit what he sees into the model of cell formation proposed by Schwann. And it is obvious that he finds this a difficult task.

In the year that Barry wrote the last of his series of papers in the *Philosophical Transactions of the Royal Society*, Carl Bergmann, then an *Assistent* of Rudolph Wagner in Göttingen and later professor of anatomy at Rostock, published a much more perceptive article in Müller's *Archiv*.[20] Bergmann's observations were made on the eggs of the brown grass frog and those of the newts, *Triton igneus* and *Triton cristatus*. He came to the clear conclusion that the furrows that eventually carved up the egg gave rise to the cells that made up the embryo:

> Nonetheless, this may be enough to show that these globules which are composed of yolk granules and contain a typical bright spot, and which now constitute the whole of the yolk, are identical with the cells from which the embryo presently develops. In the light of the correspondence between these characteristic components, this identity must now be regarded as certain.*

Although Bergmann considers that the 'bright spots' are probably nuclei, he reserves his final judgement on this point until further evidence of common properties or functions has been produced. But in a subsequent paper, published in 1842,[21] he considers that his previous vacillation was unwarranted and that the 'bright spots' are indeed nuclei. He finds it extremely difficult to accommodate his observations to the model of Schleiden and Schwann. He suggests that perhaps numerous nuclei are formed first and then distributed to the globules by the process of furrowing. However, he hesitates to accept this and cannot understand why one sees only complete cells and never the transitional forms that should be present if cells are formed *de novo*. He does not bluntly say that Schleiden and Schwann are wrong, but he allows himself the remark that their model has not been proved and must, for the time being, be regarded only as a theory.

Bergmann's 1842 paper makes many references to the work of Carl Vogt, an eminent zoologist, whose monograph on the development of the obstetric toad (*Alytes obstetricans*) was published earlier in the same year.[22] Vogt accepted Bergmann's observations on the frog, but claimed that they did not apply to *Alytes*. He took the view that, in *Alytes*, cells were formed in the yolk only after furrowing had been completed, and he suspected that this was also true for other classes. He thus regarded furrowing and cell formation as two separate events. He refers to Schwann with deep respect and has great difficulty in shaking off his doctrine. Nonetheless, he is obliged to defect from it to some extent. Vogt believed that cells could be formed without the mediation of a nucleus. Secondary cells could be entirely enveloped within other cells, and a new cell membrane could be formed at a stroke in the 'cytoblastic' fluid. In a section entitled 'Einiges über Zellen im Allgemeinen' (Some remarks on cells in general) Vogt summarizes his views on cell formation and distinguishes several alternatives, some of which contradict the model proposed by Schwann. To begin with, he asserts that the formation of the nucleolus does not necessarily precede that of the nucleus; the nucleolus may appear later. Similarly, the cell may be formed before it develops a nucleus. The first, and essential, event is, in his view, the formation of a cell membrane. In general, he admits that the cell and the nucleus develop synchronously but claims that cells may rearrange themselves by means of their membranes. Vogt did not reject the notion of a 'Cytoblastem' but thought there were two kinds: primary 'Cytoblastem' which was extracellular and composed of material that had never formed part of a cell; and secondary 'Cytoblastem' which was once composed of cells but subsequently became amorphous. Rarely, cells were said to arise within nuclei. Finally, in the larval notocord of the newt *Triturus*, Vogt saw binary fission in which the membrane bent inwards and divided the cell into two independent halves. Thus, in his efforts to accommodate Schwann but at the same time safeguard his own observations, Vogt was obliged to offer a veritable tutti-frutti of possibilities.

The work of von Siebold on invertebrates was continued by H. Bagge whose observations on the eggs of nematodes were reported in his inaugural dissertation published in Erlangen in 1841.[23] Bagge, like Bergmann, was in no doubt that the progressive segmentation of the egg gave rise to ever smaller subdivisions that eventually became the cells of the embryo. He called the products of early segmentation the *vitelli partes* (subdivisions of the yolk) and the much smaller vesicles produced by later segmentation the *globuli*. In *Ascaris acuminata*, he established the continuity of these structures, the reduction in size being produced by repeated cleavage; and, in *Strongylus auricularis*, he saw that the *vitelli partes* contained nuclei, which he called *cellulae*, thus giving rise to later confusion. In the early embryo of *Ascaris* he noted that the nuclei divided and that they did so before the division of the cell.

Heinrich Rathke, a professor of zoology and anatomy at Königsberg, whose observations on invertebrates appeared in 1842,[24] came to conclusions that did not differ greatly from those of Bagge. Working with the freshly laid eggs of the mollusc *Limnaeus*, Rathke also found that the yolk was progressively divided into a number of smaller units, and he used the word 'Durchfurchung' (cleavage) as opposed to 'Furchung' (furrowing) to indicate that the 'raspberry-like' appearance of the developing egg was not merely a surface phenomenon: 'If the yolks are now squashed, it can be seen that each of these protuberances is composed of the smaller part of a more or less spherical cell, and that indeed the whole of the yolk, including its interior, is made up entirely of cells.'* Rathke also saw nuclei in the cells and nucleoli within the nuclei, but he did not think that the nucleoli inevitably divided, for, in the snail, he found that the nuclei within the cells of the germinative region contained only a single nucleolus. Nor did he think there was much evidence for the view that nucleoli eventually generated nuclei, as Schwann had suggested. Bagge and Rathke, like Bergmann, thus had a pretty clear understanding of how the embryo developed from the fertilized egg; but the idea that embyronic cells were generated simply by repeated cleavage of blastomeres was not universally accepted.

A particularly good example of the contradictions produced by the attempt to accommodate Schwann is seen in the work of Reichert, whose ill-judged attempt to support Schwann's 'Zellentheorie' against Purkyně's 'Körnchentheorie' has already been mentioned. In 1840, the year after the publication of Schwann's monograph, Reichert's own appeared,[25] dedicated to his highly esteemed teacher, Johannes Müller. Reichert found that, in the fertilized eggs of both hen and frog, the yolk was composed entirely of cells, but the larger cells in the interior of the developing embryo had no nuclei. Instead of regarding this mistaken observation as an objection to Schwann's model, he interpreted what he saw simply as a variant of it. 'In the middle one finds larger cells without nuclei. They are concerned with the young generation of cells that are about to be produced and are usually set further back; they are mother cells at the stage when the cell nucleus is resorbed, and the young progeny is to be formed at the expense of the cell contents.'* By 1841, however, Reichert found himself forced to disagree with Schwann to some extent. In a paper dealing with the furrowing process in frogs' eggs,[26] he notes that Schwann had adumbrated that furrowing might be associated with the formation of new cells, but Reichert claims to have observed that the smallest solid components of the yolk do not change throughout the whole furrowing process and cannot therefore be involved in the generation of new cells. Moreover, he disagrees with Schwann about the role of the nucleolus. He finds that in some cases the nucleus appears before the nucleolus. He concludes that the small elementary cells ('elementar Zellen') that aggregate to form the embryo arise from the disintegration of larger cells, and that: 'The process of furrowing in the yolk of Batrachian eggs is simply the gradual progression of the act of parturition in nested mother cells that serve to produce the whole of the cellular organism.'*

Five years later, in 1846, Reichert published another paper[27] on furrowing, this time in nematodes, particularly *Strongylus auricularis*. He now admits that some of his previous observations were incorrect, but justifies himself on the grounds that he was, in his earlier work, much under the influence of Schwann. However, he still does not accept the view that new cells are formed by the cleavage of older cells. He believes that furrowing is merely a mechanism for the release of a new brood from the mother cells: 'All these phenomena, as is shown by the investigations described above, are not concerned with the formation of the spherical cells produced by furrowing, but rather with the release, the birth, out of the membranes of the mother cells, of a brood of cells that have already been formed and that still lack nuclei.'*

The confusion engendered by Schwann's monograph rumbled on for more than a decade. There is no doubt that his meticulous histology was admirable, that his extension, and indeed universalization, of the observations of Purkyně represented what some philosophers would now call a paradigm shift; but the 'general principle of cell formation' which formed what he himself regarded as the centrepiece of his work, did nothing but mislead and confuse subsequent observers. Indeed, the origin of embryonic cells from the fertilized egg was not finally settled until 1855 when Remak published his great book on the embryology of vertebrates.[28] Remak introduced the systematic use of hardening agents into animal histology. He tried a number of hardening agents, hydrochloric, sulphuric and chromic acids, mercuric chloride and alcohol, but found empirically that a mixture of 6 per cent copper sulphate and an equal volume of 20–30 per cent alcohol gave the most satisfactory results. With this reagent he was able to visualize the membrane of the egg and concluded that, in the formation of the blastoderm (the disc of cells on top of the yolk), it was this membrane that split the cell into two. He envisaged that the cleavage of the cell was brought about by a ligature that was formed by the membrane itself ('Abschnürung' again).

After Bergmann, Bagge, Rathke and Remak there was no longer any doubt about the continuity between the fertilized egg and the cells of the developing embryo. This was by no means a piece of incidental information, for this continuity provided an avenue for the transmission of the hereditary characteristics, both paternal and maternal, of the fertilized egg. The demonstration that the egg was itself a cell and that it begat daughter cells by binary fission marked a decisive step in the growth of what later became the science of genetics.

Chapter 13

Remak and Virchow

If there was one individual who, above any other, was responsible for bringing order into the confusion that shrouded the origin of animal cells, it was Robert Remak. It was not acknowledged in his lifetime, as we have seen, nor is it adequately recognized even now. Of the many modern textbooks that include a historical account of the theory that all the cells in the animal body arise from other cells, it is difficult to find one in which the discussion is not dominated by the name of Rudolf Virchow. It is to be hoped that the recent scholarly biography of Remak by Heinz-Peter Schmiedebach[1] may go some way to restoring the balance.

Remak was born in Posen in 1815 and received his early education there, but, for a variety of largely personal reasons, chose to live in Berlin for the whole of his professional life. Two-thirds of the population of Posen at that time were Poles and it was a Polish Gymnasium that Remak attended, before matriculating in Berlin in 1833. Although Posen, having been incorporated into the Prussian state, was subjected to an intense programme of Germanization, Polish national feeling remained high and found expression in a number of vigorous political movements. It is clear that Remak, despite permanent residence in Berlin, retained Polish loyalties. As in the case of Czech historians writing about Purkyně, we find that the Polish historian Wrzosek, in one of the few twentieth-century studies of the life and work of Remak, stresses his essential Polishness.[2] This did not serve him well in the conservative circles that then dominated the medical faculty in the University of Berlin. Nor did it serve him well that he participated in various liberal organizations both within the university and outside it. But the two main factors that militated against his attaining the academic career that he coveted, and to which his scientific achievements amply entitled him, were first that he was a Jew who refused to undergo baptism, and second that he had a quarrelsome disposition that was exacerbated with the passage of the years by a concatenation of personal frustrations.

The normal avenue for Jews in Germany who sought preferment in the university world was to become Christians; and even when they did, the road to a full professorship was not easy. Remak's roots in Polish orthodox Judaism and the obligations of his Jewish marriage made conversion to Christianity a moral and practical impossibility. Nonetheless, a professorship

50. Robert Remak (1815–65)

at the University of Berlin, coupled with a personal unit at the Charité, Berlin's main academic hospital, were the goals that he set himself. At the time that he embarked on his academic career, no Jew had ever attained them. Ludwig Traube, the first unconverted Jew to scale these heights, was not made an *Extraordinarius* until 1857. The Charité recruited its housemen exclusively from the Pépinière, Berlin's military medical school, and the post of *Zivilassistent* (civilian assistant) had to be created to accommodate Traube. Lest this unsavoury mixture of parochialism and anti-Semitic bias provoke modern English liberals to throw up their hands in horror, it is worth noting that no Jew was elected into a Tutorial Fellowship at an Oxford or Cambridge college until 1882, when Samuel Alexander was made a Fellow of Lincoln.[3]

Schmiedebach has examined in detail the various obstacles that Remak had to overcome and the extent to which each of them reflected the interplay of prejudice and tradition. Suffice it to say that his '*Habilitation*' (formal certification of his eligibility to teach at a university) was delayed, that he was not made an *Extraordinarius* until six years before his death, and that the adequately remunerated position of *Ordinarius* and the acquisition of a unit at the Charité remained forever beyond his grasp; and this despite the unwavering support of the highly influential Alexander von Humboldt. Remak's histological *chef d'oeuvre*, the *Untersuchungen über die Entwickelung der Wirbelthiere*[4] (Investigations on the development of vertebrates) was dedicated to Alexander von Humboldt 'in eternal gratitude for his support in a life frustrated by religious and political prejudice' (religiös–politische Unduldsamkeit). Remak's career as a microscopist was cut short by the necessity to earn an adequate living. Failure to obtain an established, salaried post in the University forced him into clinical practice where he specialized in neurology and especially galvano-therapy. His scientific contributions in this area were treated with great reserve by his contemporaries, including

Dubois-Reymond. The work lies outside the scope of the present volume and has been considered in detail by Schmiedebach.

At the end of Chapter 11 I recounted briefly how Remak came to adopt the view that all, or virtually all, the cells in the animal body arose by binary fission of pre-existing cells. And at the end of Chapter 12 I discussed his use of hardening agents to visualize the cell membrane and hence to settle, once and for all, the vexed question of blastomere formation from the fertilized egg. It is in his 1852 paper[5] that Remak sets out the position he has reached after more than a decade of preoccupation with the problem of cell generation. This paper bears the title 'Ueber extracellulare Entstehung thierischer Zellen und über Vermehrung derselben durch Theilung' (On the extracellular formation of animal cells and their multiplication by division). The title embodies the two principal conclusions that he had reached: that extracellular formation of animal cells does not exist; and that it is the general, if not universal, rule that animal cells arise by the division of pre-existing cells. The paper begins with a review of the models proposed by Schleiden and Schwann and categorically rejects them. Indeed, Remak claims to have treated them with profound scepticism from the moment they were propagated: 'As for myself, the extracellular formation of animal cells struck me, from the very moment that this theory was propagated, as no less improbable than the generatio aequivoca of organisms.'*

It is of interest that Remak at once sensed, as a modern reader does, that there was a parallel between Schwann's theory and the theory of spontaneous generation; and that anyone who rejected the latter would have difficulty with the former. Remak claims that it was his doubts about Schwann that prompted him to investigate the multiplication of red cells in the chick embryo, a claim that is not improbable as his first publications on this subject appeared in 1841,[6] a mere two years after the widespread dissemination of Schwann's monograph. He finds himself unable to confirm Reichert's observation, which will be discussed in more detail at a later stage, that the nucleus disappears before cell division and is formed anew in the daughter cells. Remak saw continuity of the nuclear material from mother to daughter cell. He refers to the experiment published in 1845[7] in which he observed binary fission of cells in primitive embryonic muscle bundles, and is now able to report that he has seen this form of cell division in many parts of the developing embryo; not, however, notably in the *chorda dorsalis*, which formed such a centre of interest for Müller and Schwann, but in the rudimentary precursor of the vertebral column. In sum, he concludes that cell multiplication by binary fission is the rule, and that this, followed by morphological modification of the new cells formed, drives the development of the whole embryo. But, being himself a pupil of Müller, he tempers his criticism of Schwann with great circumspection: 'At this point, my intention is merely to draw attention provisionally to the fact that in the primordia of the most diverse tissues one can observe the progressive division of existing cells but never the formation of extracellular nuclei or extracellular cells.'*

The concluding paragraph of the 1852 paper ends on a firmer note and extends to pathological processes the concepts that Remak has derived from embryology:

> These findings are as relevant to pathology as they are to physiology. It can hardly now be disputed that pathological tissue formations are simply variants of normal embryological patterns of differentiation, and it is not probable that it is their prerogative to generate cells in the extracellular fluid. The so-called 'organisation of plastic exudates' and the earliest stages of the development of tumours need investigation in this respect. On the strength of the confirmatory evidence generated by the scepticism that I have entertained on this subject over many years, I make bold to assert that pathological tissues are not, any more than normal tissues, formed in an extracellular cytoblastem, but are the progeny or products of normal tissues in the organism.*

The extension of these ideas to pathology and especially to malignant tumours was crucial in two respects: first, it was in direct opposition to the theory formulated by Müller himself to explain the genesis of malignant tumours; and second, it was the central theme of Virchow's *Cellularpathologie* which won Virchow fame in his own time and an honoured place in history afterwards. The origin of malignant tumours is considered in greater detail in a paper that Remak published in 1854.[8] He specifically denies that tumours are formed by extracellular formation of 'free' cells or 'free' nuclei and he now asserts without reservation that malignant tissues arise from normal ones.

In his 1855 book[9] Remak restates his position and introduces a review of his own experimental work with a detailed discussion of the cell theory as propounded by Schleiden and Schwann. Here he is quite categorical in his rejection of this theory and even expresses the view that there are two theories, both wrong. He regards Schleiden's model as essentially the endogenous formation of cells out of *intra*cellular amorphous material, whereas Schwann, unable to convince himself of this in animal cells, proposes that, in animals as opposed to plants, new cells are formed in the *inter*cellular spaces. Remak points out that this divergence implies a fundamental difference between plant and animal cells rather than a correspondence ('Übereinstimmung'), as the title of Schwann's monograph asserts. In support of his own ideas he refers to the observations of von Mohl and Nägeli on confervae, but he does not mention Dumortier. Indeed, the only reference to work outside the German tradition appears to be a cursory mention of Charles Robin. He offers cogent criticisms of the work of Reichert, Henle and Albert Kölliker, who had opposed his own interpretation of events, notably in skin and cartilage, but Remak points out that in these tissues appearances can be notoriously deceptive.

Remak reserves his sharpest criticism, however, for Schwann's analogy

51. Rudolf Virchow (1821–1902)

between cell formation and crystallization. After rehearsing the main differences between the two processes, and especially the fact that primitive precursors may differentiate into specialized cells of various types, he ends his discussion of the subject with the following words: 'It is hardly necessary to make special mention of the similarity or disparity of cells and crystals, for, in the light of the facts that I have discussed, the two structures offer no points of comparison.'* He goes on to expound how his own research makes it clear that the egg is itself a cell; he notes again that the appearances one sees in multinucleate cells may be deceptive in that they give the impression of endogenous cell formation mediated by the enclosed nuclei; and finally he commits himself unequivocally to the view that the multiplication of cells in the animal body is driven by binary fission. As far as I am aware, the discussion of the cell theory in Remak's book is the most thoroughgoing, and, in the light of modern knowledge, the most cogent review of the subject that the contemporary literature offers.

We now come to an episode that throws a curious light on one of the most notable figures in nineteenth-century medicine, Rudolf Virchow. The relationship between Virchow and Remak, which began as collegial friendship, is well known to have cooled in later years, but an article on Remak by Bruno Kisch, in a series entitled 'Forgotten Leaders in Modern Medicine'[10] accuses Virchow of outright plagiarism. The facts on which Kisch bases his case are the following. There is no doubt that Remak had by 1855 repeatedly stated it to be his view that the cells of the body multiplied by binary fission, and, beginning with his observations on red cells in 1841, he had adduced ample evidence in support of this view. There is also no doubt that Virchow was familiar with his work. Both Virchow and Remak were pupils of Müller and professional colleagues in Müller's laboratory. There is evidence that in the early stages of their careers they discussed problems and experimental findings of mutual interest. Virchow did not doubt Remak's observations on the multiplication of embryonic red cells, but he initially had serious reservations about the idea that all the cells in the animal body were formed by binary fission. It is clear from a paper that Virchow wrote in 1847[11] that he simply accepted the notion that cells were formed from a 'structureless blastema' which he regarded as a fluid composed in part of exudate from the blood vessels. By 1851[12] he had adopted the view that cells were formed within other cells, and he regarded the multinucleated cells seen in tumours as centres of cell formation. His scepticism with respect to the generality of Remak's findings is still in evidence in 1854 when he reviewed Remak's 1854 paper on tumours for Canstatt's annual report.[13] Kisch elaborates on the last sentence in that report, in which Virchow simply says that Remak insists, as he has done before, that there are no free nuclei and no extracellular formation of cells. The reserved tone of this final sentence leaves little doubt that Virchow, the reviewer, was still in two minds about this. Then, in the following year, in the *Archiv* that he himself had founded, Virchow produced a wholly unexpected leading article entitled 'Cellular-Pathologie'[14] in which he adopts, virtually without modification, the position advocated by Remak.

Kisch regards this article as plagiarism, but it can hardly be that, for Remak's papers had been published in widely distributed journals and his position was well known, although not unopposed, in medical circles. We cannot at this stage be sure, unless new primary sources come to light, about what it was that changed Virchow's mind between 1854 and 1855. It is possible that Remak's book on the embryology of vertebrates, and especially the section dealing with the cell theory, might have had a decisive influence, for even if Virchow, when he wrote his editorial, had not yet read Remak's finished volume, he must have been aware of its contents; and, at a later stage, in his own book, the celebrated *Cellularpathologie*,[15] he speaks of extending to pathology the findings that have come from embryology. So closely does Virchow follow Remak's exposition that there are even echoes in the words he uses. For example: 'According to Schwann the intercellular substance is a cytoblastema destined to produce new cells. I consider this erroneous';* or, 'In pathology also, we can go so far as to consider it a general principle that no development at all is initiated *de novo*. In the developmental history of individual components just as in the development of whole organisms, we reject the notion of *generatio aequivoca*';* or 'Nowhere is there any form of new formation other than that produced by binary fission; one element after another divides; one generation produces another.'* Nonetheless, in his editorial, Virchow fails to mention Remak.

By modern ethical standards this omission is difficult to forgive and attracted some criticism even in Virchow's time. It provoked an intemperate letter from Remak that irretrievably damaged relations between the two men. But there is a defence, and Virchow makes it in the preface to the first edition

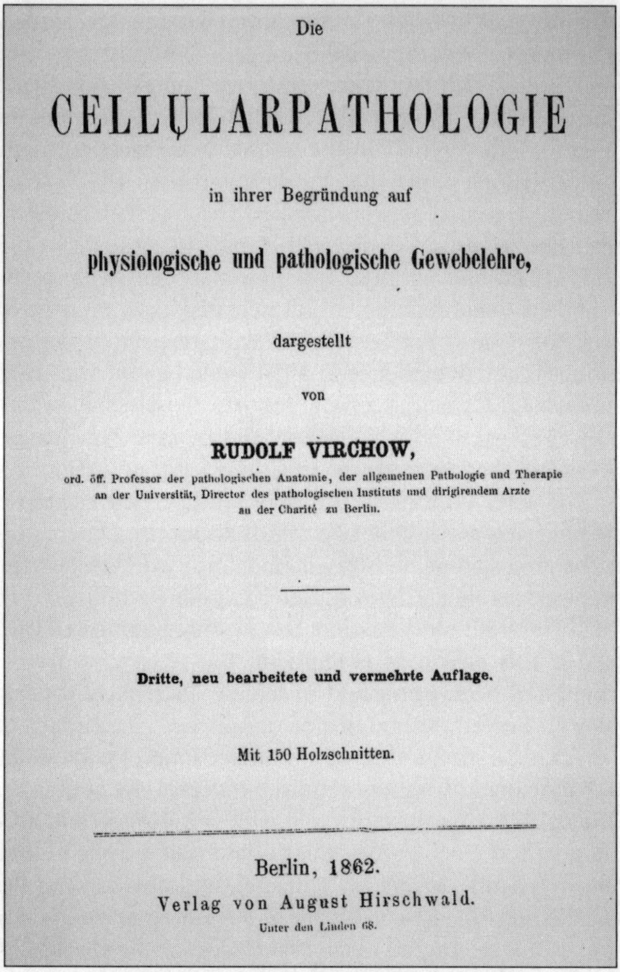

Die

CELLULARPATHOLOGIE

in ihrer Begründung auf

physiologische und pathologische Gewebelehre,

dargestellt

von

RUDOLF VIRCHOW,

ord. öff. Professor der pathologischen Anatomie, der allgemeinen Pathologie und Therapie an der Universität, Director des pathologischen Instituts und dirigirendem Arzte an der Charité zu Berlin.

Dritte, neu bearbeitete und vermehrte Auflage.

Mit 150 Holzschnitten.

Berlin, 1862.
Verlag von August Hirschwald.
Unter den Linden 68.

52. The title page of the third edition of Virchow's *Cellularpathologie*

of *Cellularpathologie*. It was not a scientific paper that he had written for his *Archiv*, but an editorial exhorting pathologists to adopt a particular point of view and a particular experimental method. So there was no need, he claims, to cite previous authors, as is the common practice in formal scientific communications: 'These days it is perhaps a merit to acknowledge historical precedence, for it is indeed astonishing how recklessly those who have found out some small detail that they extol as a discovery pass sentence on their predecessors.'* But, of course, one could hardly call Remak's contribution a 'Kleinigkeit' (small detail) even though Virchow may well have regarded him as one of several precursors who had contributed to the development of ideas about the importance of cell division.

Schmiedebach argues that Virchow deliberately chose an editorial, rather than a formal scientific paper, in which to present his views in order to avoid mentioning his immediate predecessors, and, in particular, Remak. I think this is a simplification of Virchow's motives. He seems to me to have chosen an editorial, because a formal scientific paper would hardly have had the impact he sought to achieve, especially among doctors and pathologists who were not experimental scientists. In fact he says as much: 'This defence is not the product of vain ambition nor the outcome of purely scientific endeavour. For if we wish to serve science, we have to disseminate it, not only in our own consciousness, but also in the estimation of other people.'* So, 'In an eminently practical science like medicine, at a time like ours in which experimental findings proliferate so rapidly, we have an even greater duty to make our knowledge accessible to our colleagues as a whole.'*

It is thus clear that Virchow does not lay claim to the discovery of the ideas he is putting forward, but he does lay claim to his rights ('Recht') as a propagator of these ideas. And it is in this light that we should regard both his editorial and his famous book. The latter grew out of a series of postgraduate lectures that he gave to a mixed audience in Berlin. His lectures as professor of pathology there were, of course, attended by pathologists, but his reputation and his excellent delivery also attracted adherents of other disciplines. The *Cellularpathologie* was an immediate success and, between 1858 and 1862, ran through three editions. It was quickly translated into other European languages, and it is difficult to deny that the rapid and widespread acceptance of Remak's ideas was not due to the meticulous observations of Robert Remak, but to the propagandist skill of Rudolf Virchow.

One of the factors that contributed to the durable success of Virchow's book was his use of the phrase 'Omnis cellula e cellula', a motto that is forever associated with his name. I have already commented on the prior use of these precise words by Raspail and given reasons why it is improbable that Virchow consciously plagiarized them. However, it should be pointed out that 'Omnis cellula e cellula' does not at all imply cell multiplication by binary fission. In fact, when Raspail used this phrase as an epigraph he had in mind endogenous formation of cells within existing ones, and he sought to find the intracellular organelle that was responsible for the birth of the new cell. 'Omnis cellula e cellula', as Virchow used it, is essentially a denial of spontaneous generation, which had few supporters by that time, and, more specifically, a denial of Schwann's idea that cells were generated out of an amorphous 'Cytoblastem' in the extracellular space. Cell multiplication by binary fission had to be added to 'Omnis cellula e cellula' before the position that Virchow was advocating became clear. Still, it is not uncommon, even nowadays, for a theory or a substance to become fashionable if it is given an attractive name or sobriquet. But it would be a mistake to regard *Cellularpathologie* merely as a piece of didacticism, for in it Virchow assembled a very wide range of disparate observations, some of which were his own, and imposed on them a unity that flowed from his central theme, namely, that in all organisms, plant as well as animal, the formation of tissues was driven by the binary fission of cells. In some respects Remak's relationship to Virchow resembles Purkyně's relationship to Schwann. Purkyně and Remak were the discoverers, but their voices were almost drowned in the publicity unleashed by the works of the colonizers, Schwann and Virchow.

After the publication of his book on the embryology of vertebrates, Remak returned to the study of red cells, and in 1858,[16] the year in which the first edition of *Cellularpathologie* appeared, he published yet another paper on binary fission in the erythrocytes of the chick embryo. The introduction to this paper states that the observations that he made in 1841 were confirmed by Kölliker in 1845, Gerlach in 1847 and other 'excellent histologists' including Franz Leydig, Max Schultze and, notably, Virchow; but because this work was opposed by Henle and Reichert, he took it up again in May and June of 1856 and again reached the same conclusion: that the red cells multiplied by binary fission. Moreover, although the title of the paper refers only to red cells, Remak also reviews his work on segmentation of the egg and the extrapolation of his ideas to malignant tumours.

There is no doubt that in Remak's own estimation it is he who is primarily responsible for the wholesale revision of Schwann's doctrine and that it is he who originated the theory that all animal cells, in physiological as well as pathological states, multiply by binary fission. The 1858 paper also contains some important observations on multinucleate cells, a subject on which he and Virchow disagreed. Remak noticed that if a cell in the process of division was perfused with an 0.6 per cent solution of potassium permanganate, it might not complete the process, but might instead round up again to form a binucleate or multinucleate cell. The multinucleate cells produced by aborted cell division could easily give the deceptive impression that endogenous cell formation was taking place.

Remak's interpretation was essentially the correct one, but Virchow persisted in his view that multinucleate cells were the centres of new cell formation, and he eventually prevailed on Remak to admit that this was a possibility, even a certainty. In a paper published in 1862[17] Remak wrote as follows: 'For normal tissues no unequivocal exception [to binary fission] has yet been established. But under pathological conditions the endogenous formation of cells within other cells, in accordance with the observations of Hiss, Buhl, Weber and myself, indubitably occurs.'* He recalls that in his 1852 paper he was unable to establish continuity between the nuclei of the embryo and those of the capillaries and the skin; and he thought such continuity hadn't been established either for the 'star-shaped cells, that Virchow calls connective tissue cells'.

In the end, Remak remained undecided. 'It would be one of the most interesting of discoveries if it could be shown that endogenous formation of cells and nuclei within other cells, or their equivalent, took place side by side with cell division in normal development or in the repair of damaged tissues, as well as in pathological conditions that entail the destruction of normal tissues.'* It should, however, be remembered that by 1862 Remak was no longer actively engaged in microanatomical work. In 1858 he already referred to his book as the 'Schlussheft meiner embryologischen Untersuchungen' (final volume of my embryological investigations), and in that same year he published an article on the peripheral ganglia of the intestinal nerves. This article might be thought to mark his transition to neurology. Perhaps the most important elements in the 1858 paper were Remak's observations on the division of the cell nucleus. It is probable that he saw all, or almost all, the stages of mitosis, when the chromosomes are visible, but he did not understand what they meant. What he made of mitotic figures is discussed in the next chapter.

Virchow eventually became one of the leading figures in European medicine and a notable polymath. He virtually founded the science of physical anthropology, drew up plans for the sanitation of Berlin, accompanied Schliemann on expeditions to Troy and to Egypt, and became a prominent liberal politician who consistently opposed Bismarck's policies in the Reichstag. But Schmiedebach has resuscitated a letter that throws a less attractive light on Virchow's private persona. In 1856 the University of Berlin founded a chair of pathological anatomy coupled with general pathology and therapeutics. Both Remak and Virchow were applicants. The medical faculty recommended Virchow and, after him, Remak. Actually, Remak stood very little chance. As mentioned earlier, no practising Jew had yet been appointed to a chair in Berlin, and Remak, despite his experimental brilliance, had no extensive experience in formal pathological anatomy. Virchow, on the other hand, was a product of the Pépinière, had been a *Prosektor* at the Charité and was already a highly successful professor of pathological anatomy at Würzburg. Nonetheless, Virchow regarded Remak as a serious rival.

It says a great deal about the quality of Remak's research that Virchow should have done so and that the medical faculty should have placed Remak second on their list. It is in connection with this rivalry that Virchow wrote the letter discussed by Schmiedebach. It was an appeal to his father-in-law, the gynaecologist Carl Wilhelm Meyer, to exert his influence in support of his own candidature. In particular, Virchow was worried about the role of Alexander von Humboldt, who had always shown himself a supporter of Remak. Virchow appears to have believed in some form of Jewish conspiracy theory and to have assumed that Humboldt's enthusiasm for Remak was grounded in philosemitism. Virchow's letter contains the suggestion that some member of the prominent Mendelssohn family might be approached so that a Jewish influence could be brought to bear on Humboldt in the hope that this would induce him to change his mind. In view of Virchow's notably liberal public position, it throws a remarkable light on his character to find that, when his own interests were at stake, he was not free of irrational bigotry. Years later, when his academic position was unassailable, Virchow nonetheless opposed moves to further Remak's career. It could be argued that Virchow's opposition was based on a genuinely sceptical assessment of Remak's neurological researches; but it was, to say the least, signally ungenerous.

CHAPTER 14

Division of the Cell Nucleus

It is likely that in the work that forms the subject of his two papers of 1841, Remak saw division of the cell nucleus. Since he clearly described partitioning of the nucleated embryonic chick red cell into two nucleated daughter cells, and consequently made the strongest possible case for binary fission, he could hardly have avoided the assumption that the nucleus also divided into two. But at that time, and indeed for more than a decade afterwards, there were few plausible theories, and even fewer observations, dealing with the function of the nucleus.

It is therefore not surprising that Remak did not at first have much to say about the mechanism of nuclear division. It was not until Reichert attacked Remak's conclusions and proposed an alternative model for cell multiplication that Remak turned his attention to the behaviour of the cell nucleus. Reichert published his observations in 1847 in another review for Müller's *Archiv*.[1] In a study of sperm formation in the nematode *Strongylus auricularis*, in which the sequence of developmental stages can be observed as one passes along the testis, Reichert came to the conclusion that when the cell prepared to divide, the nuclear membrane disappeared altogether and the nuclear contents were dissolved; new nuclei, he supposed, were then re-formed *de novo* in the daughter cells. The first part of this theory, namely the dissolution of the membrane surrounding the nucleus, was a discovery of obvious importance, and it proved to be correct. But the idea that the nuclear contents were also dissolved and re-formed *de novo* was to some extent due to inadequate microscopic resolution; but it was more directly an attempt to accommodate what was actually observed within the model proposed by Schleiden and Schwann to which Reichert was still in part attached. Reichert explicitly states that after the nuclear membrane disappears the cell is 'kernlos' (lacks a nucleus) and the daughter cells are also initially 'kernlos'.

By the time he came to write his 1852 paper,[2] Remak was convinced that there was continuity of the nuclear substance between mother and daughter cells, and, in this paper, he openly declared himself unable to confirm Reichert's observations. In his embryological monograph of 1855[3] he refers to his own observations on the frog's egg and repeats that right up to the late stages of segmentation the division of the cell is preceded by division of the nucleus; his findings thus leave no room for a scheme that envisages disappearance and *de novo* reappearance of the nucleus. But he tempers his remarks with a good deal of caution: 'Observations on the mechanism of nuclear division are by no means so extensive as those on the behaviour of the cell membranes.'* And he suggests that it is not yet settled whether division of the nucleus entails the dissolution of the nuclear membrane or whether appearances are deceptive just as they are in the case of the cell membrane. Actually it was difficult for Remak to conceive of any mechanism other than direct partitioning of the nucleus into two. The scheme that he proposed was that binary fission involved first the nucleolus, then the nucleus and finally the cell. His evidence for division of the nucleolus was based on counts of the number of nucleoli present before and after cell division. This seems to be one of the few examples of Remak deceiving himself in an attempt to find support for a preconceived notion, although, of course, it must be remembered that these observations long antedated the birth of modern statistical methods. Under the influence of his own work on the role of the cell membrane in the segmentation of the egg, Remak believed that the partitioning of the nucleus and nucleolus was similarly due to 'Abschnürung' produced by invagination of the relevant membrane. By the time he wrote his 1858 paper,[4] he felt pretty sure of his ground 'The rule is that the nucleolus is partitioned by the membrane into two parts and the nucleus is similarly partitioned into two nuclei.'*

Remak's self-deception was not nearly so stark as that shown by Nägeli and Virchow. Nägeli actually drew a figure in which the nucleus of one of the terminal cells in the hairs on the stamen of *Tradescantia* is shown in the process of being divided into two by means of a partition.[5] But Nägeli considered that the formation of new nuclei by the division of old ones was unusual, and his views on the generation of new nuclei were no less catholic than his views on the generation of new cells. He noted that in the spore mother cell of *Navicula striatula*, a single nucleus gave rise to two similar nuclei, one in each of the daughter cells; and in *Anthoceros* he saw in the spore mother cell the formation of four small nuclei from two nuclei. In these cases he claims to have actually observed partition of the nucleus into two: 'Now the first thing that happens is that this nucleus divides into two rounded nuclei, and then the formation of the two secondary mother cells takes place.'*

Nägeli could not have seen mitosis (the stages of cell division when the chromosomes are separated), or he would not have been able to produce a drawing showing the nucleus being divided by a partition wall. He merely assumed that this was happening and then proceeded to find examples that supported his assumption. But he also described the formation of a new nucleus within a cell while the existing nucleus was still intact and attached to the cell wall; and he obviously thought that there were many situations where the creation of the new nuclei took place *de novo*. It is also clear that he regarded the nucleus as dispensable, for he described anucleate cells and, as mentioned previously, proposed that some plant species were anucleate altogether. However, Nägeli's imaginings were restrained compared to the description of nuclear division given by Virchow. In 1857, in a paper[6] devoted to this subject, Virchow wrote as follows:

> The most common form of nuclear division takes place in the following manner. First, a small constriction or groove is formed on one side of the usually rather oval nucleus. This groove gradually extends over the surface of the nucleus, and from it the partition wall then penetrates right through the interior of the nucleus. Sometimes one sees such grooves form simultaneously at two or more sites on the periphery of the nucleus which then divides into two or more subunits. To begin with the partition wall is always completely straight, but as the nuclear subunits grow and differentiate into discrete nuclei, their boundaries also round up; and eventually the new nuclei move apart.*

Most of this is the purest fiction, but apparently Virchow has no hesitation in describing it in the most minute observational detail and devoting a whole paper to it.

In the 1840s and 1850s a number of authors thought, like Remak, that the new nuclei were sometimes formed by direct division of old ones. But few ruled out the idea that, more often, new nuclei were generated *de novo*. For example, R. Breuer, in a doctoral thesis written in 1844 on the subject of scar formation,[7] states: 'diviserunt nuclei bipartitione aut divisione multiplici' (the nuclei were partitioned into two or more parts), but adds: 'Generatio verum endogena fit aut rarius sejunctione nucleorum' ('They [the nuclei] are in fact generated endogenously, or, less frequently, by splitting of existing nuclei'). Breuer's pupil, Günzburg,[8] believed that the number of subunits into which the nucleus was partitioned was determined by the number of nucleoli that it contained. As early as 1841, the year in which Remak published his first accounts of cell division in embryonic erythrocytes, Bagge[9] noted that in the early embryo of *Ascaris nigrovenosa* division of the nucleus preceded division of the cell, but he says little about the mechanism of the nuclear division. Carl Gegenbaur, one of the most eminent comparative anatomists in Germany, whose paper on the development of *Sagitta*,[10] appeared one year after Virchow's elaborate account of nuclear division, is much more circumspect in what he says:

> How it [the nucleus] behaves during division has at the critical moment escaped me, but it should be noted that I frequently saw a stage at which the nuclei were greatly elongated. Many of them also had constrictions so that, without actually having seen a divided nucleus, I can nonetheless conclude that division was taking place. In addition, in no case did I see a cell without a nucleus.*

This lengthening of the nucleus was seen by several authors and might well have been a poorly resolved mitosis in which the chromosomes were to some extent visible, or a spindle (the fusiform array of fibres that guides the movement of the chromosomes), or both.

Kölliker, who had been Gegenbaur's teacher and was now professor of physiology and comparative anatomy in Zürich, writing about the embryology of cephalopods,[11] mentions lengthening of the nucleus but claims to have seen it split into two: '. . . the multiplication of nuclei, which is the only way in which nuclei are produced for the generations of cells that are formed endogenously within the primary cells, took place in the following manner. I saw them elongate, become constricted across their middle and finally divide into two.'* However, this passage is preceded by a recitation of Schwann's views, in which Kölliker acquiesces declaring that crystallization of the nuclei from a homogeneous fluid is the most probable course of events ('höchst wahrscheinlich'). Von Baer, writing from St Petersburg in 1846,[12] also describes lengthening of the nucleus in the egg of *Echinus lividus*, but his description is more graphic than that of Kölliker. The ends of the nucleus are said to swell up and the middle to contract; and when the nucleus has divided into two, the swollen ends, which now have trailing appendages, revert to their spherical form. Max Schultze, of whom more presently, concluded in 1861[13] that the nuclei of muscle fibres arose by partitioning of existing nuclei, and August Weismann,[14] in 1863, thought that nuclear division occurred at the same time as cell division in the early development of *Musca vomitoria*.

But there were still those who adhered to Reichert's doctrine that the nucleus disappeared at each cell division. For example, Krohn,[15] writing in 1852 about the development of ascidians, has this to say: 'As far as the bright vesicular nuclei within the blastomeres are concerned, I believe that I can conclude that at each impending cell division, they disappear and are reformed only when the division of the cell has been completed.'* Even as late as 1870 Johannes von Hannstein, a professor of botany in Bonn, was still disputing Reichert's doctrine.[16] Hannstein believed that the nucleus was bisected by a delicate ligature ('Halbierungsgrenze'), that the two halves then moved apart, and that a new cell wall was formed between them.

It is not easy to identify who first discerned chromosomes during mitosis, but there is no doubt that those who first saw them had no idea of their significance. Henle in his textbook of 1841[17] describes the elongation of nuclei and then their transformation into slender strands. The nucleolus is said to disappear and be replaced by small granules. Baker considers that Henle's claim is very thin and allocates the primacy to Nägeli. In a monograph written in 1842,[18] Nägeli describes how the cytoblast (nucleus) in *Lilium* and *Tradescantia* is replaced by a number of much smaller, transitory cytoblasts. A plate reproducing these smaller cytoblasts accompanies the description, but whether they are chromosomes is in the eye of the beholder. The first unequivocal illustration of chromosomes is, according to Baker, that given by Alexander Kowalevski, writing in 1871[19] from Kiev where he was then professor of zoology. But this is surely an error on Baker's part, or at the least a rather special interpretation of what one chooses to call a chromosome. For

53. Wilhelm Hofmeister (1824–77)

Baker's conclusion overlooks the illustrations accompanying Remak's paper of 1858[20] and those of Wilhelm Hofmeister which were published in 1848[21] and 1849.[22] Remak's plates show recognizable anaphase and telophase figures (the subdivisions of mitosis during which the chromosomes can be seen to be separating into two groups and those during which they have separated completely), but he describes them as 'verschrumpfte Kerne' (shrivelled nuclei) and he presumably regarded them as artefacts. Remak was at the time convinced that both nuclei and nucleoli split into two, and he clearly had no idea of the biological significance of the appearances that he accurately reproduced.

Hofmeister, working with plant material, was able to identify, and illustrate, virtually all the stages of mitosis that we now recognize. In the series of three papers written in 1848, he reported that in the pollen mother cells and staminal hairs of *Tradescantia* the nuclear membrane dissolved before cell division, but he thought that the contents of the nuclei were still present. When the cells were stained with iodine, 'Klumpen' (discrete lumps) derived from the nuclei were still visible. Hofmeister believed that these might correspond to the small transitory cytoblasts that Nägeli had described. 'Granular mucilage' appeared to collect around these lumps, and when the cell divided the aggregate of nuclear material separated into two masses. These received a membrane, and thus formed the nuclei of the daughter cells. Both the papers of 1848[23] and the monograph of 1849[24] are accompanied by plates, which, among other things, illustrate with remarkable accuracy the following mitotic events, first in *Tradescantia* and later in *Passiflora coerulea* and *Pinus maritima*: visible morphological changes in the cell nucleus before cell division; dissolution of the nuclear membrane; unmistakeable chromosomes attached to a metaphase plate (an equatorial plate formed in the divid-

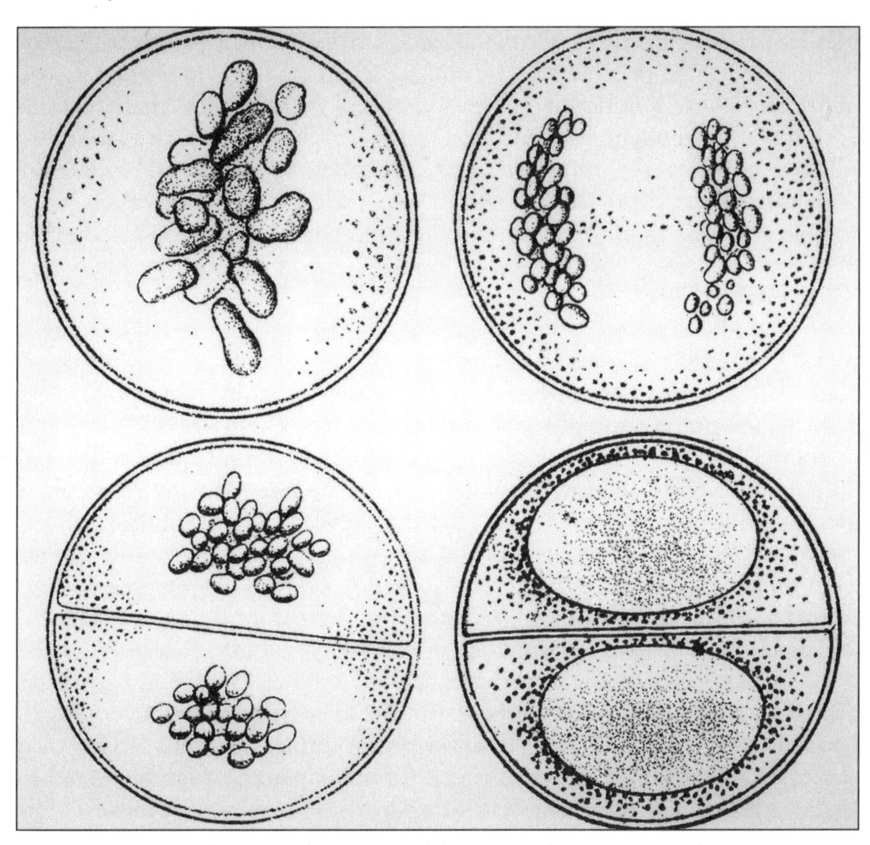

54. Hofmeister's illustration of the stages of mitosis: metaphase (upper left); anaphase (upper right); telophase (lower left); reconstitution of the daughter nuclei (lower right)

ing cell); anaphase; reconstitution of the nuclei in the daughter cells, initially without a membrane; formation of a nuclear membrane around the newly reconstituted nuclei; formation of a partition wall (Scheidewand) between the two daughter cells. These appearances were also seen when the daughter cells divided and groups of four cells were formed.

Hofmeister did not, of course, have any conception of the biological role of the phenomena that he described, but he was fully aware that they were not chance events. He suggests that the 'Klumpen' generated by the dividing nucleus were coagulates of protein, but by this he is unlikely to have meant a random process like protein precipitation. In a textbook published in 1867,[25] he still adhered to the view that the 'Klumpen' were protein coagulates, but in view of their regular recurrence he cannot have avoided the conclusion that they were produced by an organized process. Hofmeister's book contains an accurate illustration of meiosis (division of the cell nucleus to produce four daughter nuclei, each with one half of the chromosomes present in the original nucleus) in a species of *Psilotum*, and his estimate of the number of lumps generated by nuclei dividing in this way was very close to one half of the number of chromosomes that we now know this organism to contain. Although nuclear partitioning, as envisaged by Remak and Nägeli, still had its adherents, the meticulous and extensive studies of Hofmeister consolidated the idea that division of the nucleus involved a much more complex process than this. The term 'indirect' nuclear division to describe this more complex process appears to have been introduced by Flemming but not until 1879.[26]

As might be expected from the state of microscopy in the mid-nineteenth century, structural studies on animal cells continued to lag behind those on plant cells. But this does not explain the passage of more than twenty years between the publication of Hofmeister's observations and the identification in animal cells of structures corresponding to the mitotic figures he had described. Despite the fact that many of the botanists at that time had a medical education, it seems that there was some degree of intellectual dissociation between the microscopical study of animal tissues and similar work on plants. The investigations of Kowalevski (1871)[27] were mainly concerned with the embryology of nematodes and arthropods: *Sagitta*, *Euaxes*, the annelid *Lumbricus*, *Hydrophilus*, *Apis* and a few butterflies. One of the figures accompanying his paper does indeed show a spindle and what are probably chromosomes in anaphase, but the illustration given is very small. It was Russow, again a botanist, who saw the connection between his own observations and those of Hofmeister. Russow's paper, which appeared in 1872,[28] deals with the formation of spores in vascular cryptogams. His figures are also small and the components illustrated difficult to distinguish in detail. It is, however, quite clear that he saw equatorial plates with chromosomes attached to them. He refers to the plate as a 'Stäbchenplatte' and to the chromosomes themselves as 'Stäbchen' (small rods). These appearances were seen in the spore mother cells of a number of cryptogams, but the large chromosomes of *Lilium bulbiferum* yielded the clearest preparations. Russow describes the equatorial plate as composed 'aus länglichen Körnchen oder kurzen Stäbchen, die hell und stark lichtbrechend sind' (of elongated granules or short rods that are bright and highly refractive). He refers to Hofmeister's illustrations and acknowledges the identity of what they both saw, but he takes issue with Hofmeister's view that the mitotic figures were protein coagulates produced by long exposure of the plant to water. Russow argues that his 'Stäbchen' and 'Stäbchenplatte' cannot be artefacts for three reasons: first, they occur reproducibly; second, the plate is always to be found in the same place and in the same orientation within the cell; and third, similar appearances are seen in intact cells, in the sporangia of *Polypodum vulgare* and *Aspidium felix*. The mitotic figures he describes are not, therefore, random protein precipitates although, as mentioned earlier, it is not clear that random precipitates are what Hofmeister had in mind.

Anton Schneider (1873),[29] at that time a professor of zoology in Giessen and a former *Assistent* of Johannes Müller, described 'dicke Stangen' (thick sticks or rods) that appeared during cleavage of the eggs of the tubellarian worm, *Mesostomum ehrenbergii*. He saw similar appearances at later stages in the development of the embryo and noted that when the cell divided half of the 'Stangen' went to one pole and half to the other. He believed that this might be the fate of the nuclei that were thought to disappear completely during cell division; but he also believed that 'direct' partitioning of the nucleus occurred in some cases.

Tschistiakoff's paper, which appeared as a series of communications to the *Botanische Zeitung* in 1875,[30] begins with a criticism of Russow for having ignored Tschistiakoff's previous work. It is an indication of the extent of German influence in Russian academic life at the time that Kowalevski, writing from Kiev, Tschistiakoff from Moscow, and Russow, whose paper appeared in the journal of the Imperial Academy of Sciences of St Petersburg, all wrote in German. The studies of Tschistiakoff dealt with the formation of spores and pollen in a variety of plants including ferns. In the spore mother cells of *Isoetes durieui* he found that, when the cell divided, a clearly defined equatorial plate, which he called 'Theilungslamelle', appeared, always in the same position. The figure illustrating this plate also shows a spindle, and bodies at the poles. Tschistiakoff believed, however, that the new nuclei were formed at the poles of the cell while the chromosomes were still at the equatorial plate. This is, of course, simply a misinterpretation of the observations which were otherwise very accurate. Of pollen mother cells he writes:

> Somewhat later both the pronucleolus and the pronucleus show quite characteristic striations; on their surfaces there appear numerous vermiform lines made of dense, refractile, plasmatic material. These turn out to be differentiation lines in the two pronuclei. In the following stages these lines are converted into broad, very dense, refractile filaments which form prominent meridians and only become visible through the action of water. The metaphase plate [the stage of mitosis when the chromosomes are aligned along the plate] of the pronucleus is also extended; it assumes an equatorial position and is made up of thickish, refractile plasmatic 'Klumpen' [he uses the same word as Hofmeister]. At about this time at the two poles where the meridian filaments converge, one notices the appearance of two plasmatic regions which are initially less dense and which later give rise to two daughter nuclei.*

It should be noted that Tschistiakoff not only uses Hofmeister's term 'Klumpen', he also claims that the spindle only becomes visible when the cell is treated with water, which Hofmeister thought to be responsible for the formation of his protein coagulates.

Baker credits von Ewetsky (1875)[31] with the first clear illustration of prophase (the earliest stage in the division of the nucleus). (Despite the Slav name, and the German title, he wrote from the Pathological Institute in Zürich.) There is no doubt that his is an excellent representation of this phase of mitosis, but it should not be forgotten that Hofmeister, more than a quarter of a century earlier, saw and illustrated morphological changes in the nucleus prior to cell division, and these were clearly depictions of prophase. Von Ewetsky describes prophase in the following terms: 'From the fifth day onward, in the usual endothelial layers, one often sees cells whose nuclei contain a few or sometimes a great many elongated, refractile bodies having the shape of rods or filaments. These are frequently tightly wound together or entwined and, indeed, form veritable tangles of threads.'* Nonetheless, von Ewetsky believed in direct partitioning of the nucleus into two or more daughter nuclei.

It was not until 1875 that the connection was made between mitotic figures in animal cells and those in plants. In his influential book *Zellbildung und Zelltheilung* (Cell formation and cell division) (1875),[32] Eduard Strasburger stressed the homology of nuclear division in the two kingdoms. This apparently dawned on him when he examined the work of Otto Bütschli. Bütschli was at the time studying cleavage and polar body formation in the nematode *Cucullanus elegans* and saw both metaphases and anaphases. He noticed that when the nucleus disappeared a structure that we now know to be the spindle was formed, and at its equator there appeared an aggregate of granules which he thought to be swellings in the spindle fibres. Bütschli observed that this single equatorial aggregate of granules gave rise to two, and that these in due course moved apart to the poles of the cell. The first account of these observations appeared in 1875.[33] Strasburger, who had been working with a variety of plant and animal material, especially on the fertilization of conifers and on mitosis in mammalian cartilage, wrote to Bütschli who generously sent him as yet unpublished illustrations of mitosis in *Cucullanus elegans* and meiosis in *Blatta germanica*. These illustrations were incorporated and acknowledged in Strasburger's book. They were not published by Bütschli until 1876.[34] Both Bütschli and Strasburger clearly observed metaphase and anaphase figures but were unsure about telophase and interpreted the role of the spindle in different ways. Strasburger thought that the spindle itself with its equatorial plate was the nucleus and that the daughter nuclei were formed by fusion of the products of the divided plate with the two halves of the spindle. Bütschli, on the other hand, noticed the polar bodies (Richtungsbläschen) forming at the ends of the spindle and, as mentioned earlier, regarded the 'dunkle Stäbchen' (dark rodlets) as swellings of the spindle fibres. Strasburger also studied mitosis in living cells, choosing *Spirogyra* as the most suitable experimental object for this purpose. Bütschli, interested primarily in animal cells, extended to embryonic chick erythrocytes the observations he had made on nematodes. He was thus able to give an accurate description of mitosis in the very material that had led Remak to the erroneous assumption that nuclear division took place by 'direct' partitioning.

55. Eduard Strasburger (1844–1912)

Bütschli's work on the fertilization of the egg will be considered at a later stage. It was begun shortly after the Franco-Prussian war when he was twenty-four years old and had moved to Kiel to take up the post of *Assistent* to Karl Möbius, the professor of zoology there. Bütschli, however, disliked Kiel and, after a few years, returned to his native Frankfurt. In 1878 he accepted an invitation to the professorship of zoology at Heidelberg where he remained for the rest of his life. Strasburger, although originally from

56. Otto Bütschli (1848–1920)

Warsaw, was the professor of botany at Jena from 1869 to 1881 and at Bonn from 1881 till his death in 1912. It is thus not surprising that Bütschli and Strasburger were much more influential, as judged by citations of their work, than the Russian cytologists, Russow and Tschistiakoff, although it is certainly the case for Russow and probably so for Tschistiakoff that their work antedated that of Bütschli and Strasburger.

The appearance of Strasburger's book in 1875 and Bütschli's long paper in 1876 prompted other authors to offer similar descriptions of mitosis in a range of different organisms. Among the earliest were W. Mayzel(1875)[35] and C. J. Eberth (1876).[36] Mayzel, writing from Warsaw, sent in a preliminary communication in which he pointed out that he had for some time been studying epithelial regeneration in the frog and had noticed the curious nuclei of cells undergoing division, but it was not until he had read the papers by Strasburger and Bütschli that he came to appreciate the significance of what he saw. Mayzel submitted another brief communication in 1877[37] pointing out that the work had been reported in 1876, first in Polish in the *Gazeta lekarska* (Medical gazette) and then in Russian at a meeting of the Society of Russian Biologists and Doctors held in Warsaw in that year. Mayzel's second communication contains an extension of his work on epithelial regeneration to many other species, including birds, mammals and man. His descriptions of mitosis dealt mainly with metaphase and anaphase as described by Bütschli and Strasburger. What appears to be the spindle is described as '. . . bisquitförmig oder in Gestalt eines aus Fasern zusammengesetzen Stundenglas' (. . . biscuit-shaped or resembling an hour-glass made up of fibres). In his 1876 paper, Carl Eberth, then a professor of pathology in Zürich, quotes Bütschli, Strasburger and von Ewetsky. He, too, claimed that he had already been studying epithelial regeneration in 1873 and 1874 but had not appreciated the significance of the mitotic figures he had observed. Now, however, he was convinced that they were analogous to what Strasburger had described in plants. The figures accompanying Eberth's paper show essentially all the stages of mitosis in both epithelium and endothelium undergoing regeneration.

Until 1876 virtually all investigations of mitosis were of a descriptive nature and were designed to establish the details of 'indirect' nuclear division. But there was no glimmer of any theory that might explain why it was that nuclei had to divide in this peculiarly elaborate way. The first step in an experimental programme that was eventually to provide a satisfactory answer to this question was the report communicated to the Académie des sciences by Edouard Balbiani in 1876.[38]

CHAPTER 15

The Indispensability of the Cell Membrane

Whether cells were inevitably bounded by a membrane was a question that arose because the solid walls that characterized plant cells could not with certainty be identified in animal cells. Despite the correspondence between the two asserted by Schwann, this was one respect in which they differed. Several real structures, such as the vitelline membrane of eggs, and numerous optical artefacts were canvassed as the animal cell analogues of plant cell walls; but, on the whole, these analogies were not accepted. Wherever animal cells were seen to move, this movement was itself thought to be an argument against the presence of solid cell walls, as, for example, the movements of frog blastomeres that Ecker[1] claimed to have noticed. On the other hand, Reichert (1841),[2] also examining the development of the egg in amphibia, found that in the presence of distilled water a membrane was detached from the surface of the cell by a process of endosmosis. As discussed previously, Remak used hardening agents to demonstrate the membranes enclosing the frog's egg. Indeed, his explanation of cleavage in the egg, and of cell division in general, rested on the activity of the cell membrane in forming a ligature to divide the cell into two. While there was strong resistance to his view that binary fission was the mechanism responsible for virtually all cell multiplication, there were few who doubted that some kind of membrane was involved in binary fission, where this occurred.

Objections to the idea that animal cells were enclosed by a membrane probably had their origins in the observed movements of some protozoa and in the multinuclear character of some primitive slime moulds. It is clear that some workers used the word membrane (Membran) to connote a rigid structure like the cell wall of plants. Since cell movement or cell locomotion was thought *ipso facto* to exclude the presence of a 'Membran', it appears that for some investigators this term did not, at least initially, connote a flexible structure. Remak used the demotic word 'Hülle' (coat or covering) as an alternative to 'Membran', thus indicating that for him 'Membran' did not necessarily imply rigidity, but there were those for whom it clearly did.

Anton de Bary, successively professor of botany at Freiburg, Halle and Strasburg, and strongly influenced by his own work on slime moulds, declared in 1860 that the swarmers had no cell membrane as the word was normally used. For de Bary,[3] the normal usage of the word apparently signified a solid

57. Max Schultze (1825–74)

structure like the plant cell wall. The most primitive animal cells, he claimed, had no skins. Others might make them or not. The other factor that greatly influenced de Bary's thinking on this question was the fact that slime moulds, which he had studied for many years and which formed the subject of his classic monograph,[4] were often multinuclear plasmodia or assumed this state at some point in their life cycle. The presence of many nuclei enclosed within one protoplasmic mass that had no visible internal partitions naturally suggested that the fundamental cellular unit was a nucleus surrounded by naked cytoplasm and not necessarily bounded by a membrane.

De Bary's views were confirmed by Max Schultze, professor of botany in Bonn and the main proponent of the membraneless cell. In 1858 he published a paper[5] that included observations on cytoplasmic movements in a range of North Sea diatoms and reached the conclusion that they were analogous to those seen by Unger and Cohn in the cells of other organisms, including higher plants. De Bary had no doubt that these movements were due to the intrinsic contractility of protoplasm and were not driven by external forces. The position Schultze had reached by 1860[6] was that the cell was originally membraneless and that structures resembling membranes were artefacts produced by the hardening process. Again, it was the argument that a 'Membran' would inhibit cell movement that Schultze found most convincing, indeed hardly contestable:

> Given the contractility of the protoplasm, alterations in the shape of the whole cell are naturally impeded, if not altogether rendered impossible, by the presence of a rigid cell membrane. The less the surface of the proto-

> plasm is hardened into a membrane, the nearer the cell approaches the original membraneless state in which it is no more than a naked blob of protoplasm containing a nucleus, and the less constrained and inhibited are its movements.*

Schultze's paper of 1861[7] is concerned with muscle and bears the alternative title 'und das was man eine Zelle zu nennen habe' (and what one should call a cell). He refers specifically to the work of Remak on hardening agents and declares that although he was indeed able to repeat Remak's observations on the developing frog's egg, later experiments of his own raised doubts in his mind. He was still convinced that the membrane visualized by hardening agents was an artefact and that during cleavage of the egg no chemically distinct membrane was present. The archetypal cell is finally defined as a nucleated blob of protoplasm that lacks any kind of coat ('hüllenlos'). Like de Bary in the case of slime moulds, Schultze regarded the multinucleated state of muscle fibres as a telling reinforcement of his views. These were not generally accepted, but they remained a bone of contention for many years.

Ernst Haeckel, who was one of the earliest investigators to argue for the centrality of the nucleus in the life of the cell, none the less considered that there existed an anucleate protoplasmic entity that he named a 'cytode' and that was not necessarily bounded by a membrane.[8] In the case of the radiolaria, about which he wrote the standard text,[9] he was convinced that cell movement was mediated entirely by an extracapsular layer of protoplasm: 'Motility in the Radiolaria is thus attributable exclusively to the extracapsular region of the sarcode.'* As late as 1880, von Hannstein[10] proposed that

58. Hugo de Vries (1848–1935)

the basic unit be called a protoplast, and that this might or might not secrete a wall. He noted, however, that in plants there was a more rigid outer skin which he called 'äussere Hautschicht' (outer integument) and a more flexible inner skin which he called 'innere Hautschicht'. This might have been the first glimmer of recognition that there was both a cell wall and a cell membrane and that the two were different.

It was again the botanists who provided the decisive evidence. Hugo de Vries, professor of botany in Amsterdam and, at a later stage, one of the trio who resurrected the work of Gregor Mendel, not only made it clear in 1884[11] that there was a membrane quite distinct from the cell wall, he also devised systematic procedures for studying the properties of this membrane. In his paper, which is concerned mainly with the measurement of turgor pressure, de Vries derives an 'isotonic coefficient' and gives it the following definition: '. . . this measures the magnitude of the affinity of a molecule of the substance in question for water in a dilute aqueous solution'.* The methodology introduced by de Vries rests essentially on the movement of water and salts into or out of the plant cell when it is suspended in various salt solutions. This is, of course, the endosmosis and exosmosis of Dutrochet, but de Vries mentions Dutrochet only in a footnote. He does, however, discuss the observation made by Nathanael Pringsheim thirty years earlier that, in dilute solutions, the contents of the cell fall away from the cell wall.[12] De Vries names this phenomenon plasmolysis: 'If a fully grown cell is suspended in a strong salt solution, it is well known that the layer of living cytoplasm is released from the cell wall and retracts into a smaller volume. This is due to the fact that the cell sap that it encloses releases water into the surrounding solution. The weaker this solution is, the smaller the contraction or plasmolysis.'*

De Vries also refers to the work of Nägeli, published in 1855, which actually includes the words 'endosmosis' and 'exosmosis' in the title.[13] In a wide range of higher plants, de Vries studied the endosmosis and exosmosis produced by various salt solutions, especially those containing potassium and calcium. He determined, in addition, that there was a relationship between the isotonic coefficient of a solution and the lowering of its freezing point, and he was fully aware that the osmotic phenomena that he was measuring were mediated by the membrane that surrounded the vacuole of the plant cell and not by its rigid cell wall. The distinction between the plant cell wall and the cytoplasmic membrane was finally established by the classical studies of Ernest Overton (1895, 1899, 1900).[14] Overton worked mainly with *Spirogyra* and used plasmolysis in this organism to provide decisive evidence that it was the 'Grenzschicht' (boundary layer) of the cytoplasm, and not the cellulose wall of the plant, that was responsible for osmosis. Overton also showed that it was the lipid solubility of a substance that determined whether it penetrated the cell membrane or not. It was a long time before work of comparable precision was done on animal cells, but, by the turn of the century, no one doubted that animal cells were bounded by a membrane, even if, unlike plant cells, they had no cellulose wall.

CHAPTER 16

Chromosomes

With the work of Balbiani and van Beneden we move away from debate about the mechanism of nuclear division to a precise delineation of chromosomes and what they do during division of the cell. Edouard-Gérard Balbiani was born in Haiti in 1823. His father was of Italian origin and his mother a French Creole. He obtained his biological training in France, and at the age of thirty-one became professor of embryogeny at the Collège de France, a post that he held for the rest of his life. In a series of papers that appeared in 1861[1] and dealt with sexual reproduction in protozoa, Balbiani described metaphase and probably prophase in the micronuclei of paramecia. His illustrations clearly show several metaphase plates, but, perhaps under the influence of Ehrenberg, he misinterpreted what he saw, and thought that the metaphase plates were the testicles of the organism and the macronucleus the ovary. This imaginative foray was corrected by Bütschli (1876)[2] who showed that the structures observed by Balbiani were division products of the micronucleus. By 1876,[3] Balbiani's observations had reached an altogether different level of precision. Studying the epithelium of the ovary in the grasshopper *Stenobothrus*, which was particularly suitable material for cytological work, Balbiani saw essentially all the stages of mitosis and noted that, when the cell divided, the nucleus dissolved into a collection of 'bâtonnets étroits' (narrow little rods). The term 'bâtonnet' to describe these structures had been used in the previous year[4] by Edouard van Beneden in his account of nuclear duplication in the ectoderm of the rabbit embryo.

Van Beneden, whose work will be considered in some detail later, recognised that the 'bâtonnets' at the equatorial plate separated into two groups which formed what he called 'disques nucléaires' (nuclear discs). These then moved apart and expanded to form the daughter nuclei. But Balbiani made two additional observations: first, that the 'bâtonnets' were of unequal sizes, and second, that each of them divided into two. Balbiani does not give a precise description of the manner in which the 'bâtonnets' are divided, but the impression given in his paper is that he thought the division occurred transversely, across the middle of each 'bâtonnet'. Certainly, he did not envisage that the 'bâtonnets' were divided into two by longitudinal splitting. The products of the division were then seen to be collected into two groups that moved apart and eventually gave rise to the daughter nuclei in which the 'bâtonnets' were

59. Edouard Balbiani (1825–99)

once again fused together. The significance of Balbiani's observation that the 'bâtonnets' were of different sizes did not become fully apparent until Sutton published his paper in 1902,[5] but it was in any case clear that the 'bâtonnets' were not identical. And the observation that each 'bâtonnet' divided into two provided the beginnings of an explanation for the fact that division of the nucleus was 'indirect'. It was 'indirect' in order to achieve precise partitioning of the nuclear contents into two qualitatively equivalent halves.

Strasburger continued to produce papers that were an egregious mixture of dedicated observation and intransigent misinterpretation. In the first edition of *Zellbildung und Zelltheilung*, which appeared in 1875,[6] he argued that the nuclei of the endosperm were formed *de novo* essentially as Schleiden had proposed. In 1877[7] he declared that he was, on the whole, in agreement with Eberth but insisted that he had never seen the cell nucleus divide into more than two pieces. There was at the time a good deal of controversy about the nature and formation of the spindle. Strasburger stressed that, in the light of his own observations and the preparations that had been sent to him from Warsaw by Mayzel, he concluded that the spindle was formed by conversion of the whole of the protoplasmic components of the nucleus ('protoplasmatische Kernsubstanz').

Since publishing his original description of mitosis, Mayzel had adopted the technique of suspending his cells in aqueous humour, and he was thus able to see, in the living state, what he had previously observed only in fixed preparations. The preparations that he sent to Strasburger were particularly good examples of spindle formation in cells from the newt *Triton cristatus*. Strasburger himself found the cells in the staminal hairs of *Tradescantia* to be

60. Walther Flemming (1843–1905)

the most suitable material for the study of mitosis and in these cells he observed virtually all the stages of the process in the living state. These he described in detail in a paper that appeared in 1879.[8] However, in the third edition of his *Zellbildung und Zelltheilung*, published in 1880,[9] he asserted that the metaphase plate itself divided into two, that the chromosomal rods were normally, but not always, aligned along the spindle and that their partition was entirely arbitrary. If a rod happened to straddle the line of bisection of the metaphase plate, then it was divided at that point. Each divided rod moved to the pole that was closer to it. Although Walther Flemming, then professor of anatomy in Kiel, gave his first lecture on the longitudinal splitting of the rods in 1878[10] and published two papers on the details of 'indirect' nuclear division in 1879,[11,12] Strasburger insisted in 1880 that longitudinal splitting never occurred except when a rod happened to lie in the appropriate plane at the metaphase plate.

Flemming claimed that it was on the very day that he gave his lecture in Kiel (1 August 1878) that he read Peremeschko's paper.[13] Peremeschko had sent a brief report from Kiev describing the observations he had made on the cells in the transparent tail of *Triton cristatus*. He found that, in epithelial cells, stellate connective tissue cells, white blood corpuscles and endothelial cells, division of the nucleus occurred in the same way. The whole cell appeared to fill with granules and 'Fäden' (threads) which in due course formed the new nuclei: 'As soon as the threads appear, those in the middle of the barrel become slightly thicker (the thickenings do not usually lie in the same plane); then the threads in these thickenings tear and the barrel divides into two equal parts which at once separate. That is how the two new nuclei are formed.'* Peremeschko states that he had already written his report when he read Schleicher's paper on the division of cartilage cells. The paper to which

Peremeschko refers appeared in 1874;[14] a second paper, also on cartilage cells, appeared in 1879.[15] Schleicher demonstrated the whole process of mitosis, including the reconstitution of the daughter nuclei, in slices of amphibian embryo cartilage and the scapular cartilage of young frogs and newborn kittens. Schleicher gave the name 'karyokinesis' to the whole process of 'indirect' nuclear division, and Peremeschko admits that his own observations differ from those of Schleicher only in detail. The fact that Flemming claimed to have read Peremeschko's paper on the very day he first reported his own findings, and Peremeschko claimed to have written his report before he read Schleicher's paper, indicates how closely similar the work of all three men was, and, perhaps, how competitive they were. However, although both Schleicher and Peremeschko observed that the chromosomal 'threads' divided, there is nothing in their papers to indicated they were aware that the 'threads' split longitudinally. That discovery belongs to Flemming.

Flemming's papers make laborious reading. Whereas Peremeschko is content to say that his work differs from Schleicher's only in details, it is precisely on the details that distinguish his own work from that of others that Flemming concentrates. Of course, the longitudinal splitting of the chromosomes was not a detail, and Strasburger's refusal to admit this attracted an excessive dose of polemic which, to my mind, does not enhance Flemming's papers. Nonetheless, there is no doubt that these papers, published over a number of years, provided the definitive account of mitosis as it was revealed by the microscopes of the day. Flemming intially chose the salamander, *Salamandra maculata*, as his experimental material, largely because this animal has large nuclei within large cells. His paper of 1877[16] is limited to observations on the interphase nucleus and has a lengthy section in which his own and earlier observations are critically compared. Flemming explored numerous fixatives and stains and also examined mesentery, lung and bladder wall in the living state by taking these transparent tissues directly to the microscope stage. The paper that he read in Kiel in 1878[17] still deals with the cells of the salamander, but now a much wider range of tissues is examined, some in both the fixed and the living state. The initial phases of mitosis (Anfangsphasen), in which the 'Fäden' (threads) are formed from the substance of the interphase nucleus, are clearly defined. The chromosomes at the metaphase plate are described as a 'Stern' (star), and, since the nucleolus disappears when the chromosomes thicken, the assumption is made that it supplies the additional material for this purpose. Flemming stresses the fact that the chromosomes split longitudinally, but he does not yet see that the two halves of the divided chromosome go to opposite poles. This paper contains the first of his many criticisms of the scheme proposed by Strasburger.

In 1879 Flemming wrote one paper for Virchow's *Archiv*[18] and one for the *Archiv für microskopische Anatomie*.[19] The first of these is essentially a criticism of the notion that the nucleus is divided by simple scission. He refers to the observations of Virchow, the thesis submitted to the University of Erlangen by A. Heller in 1869, and to the work of Remak, all of whom considered that 'direct' division of the nucleus was the norm. In particular he is critical of the scheme proposed by Remak according to which first the nucleolus, then the nucleus, and finally the cell all undergo 'direct' partitioning. Flemming claimed to have studied such appearances for five years and come to the conclusion that they could readily be interpreted as evidence for 'indirect' nuclear division. He does, however, agree with Remak about the mode of formation of multinucleate cells. These, he claims, arise from nuclear division (but indirect) within a cell in which the act of division does not go to completion. Only in motile cells such as leucocytes does he admit that direct nuclear fission might possibly occur, but, as far as he is aware, this has never actually been observed. The paper submitted to the *Archiv für microskopische Anatomie* in 1879 is the first part of a series in which the anatomy of cell division is described in minute detail. Flemming naturally elaborates on the 'Längsspaltung' (longitudinal splitting) of the chromosomes and suggests, tentatively, that one half of each split chromosome might go to each daughter nucleus. He also gives it as his view that the chromosomes are bound together end to end in a continuous skein (Knäuel). To denote the conversion of the nucleus into threads Flemming uses Schleicher's term 'Karyokinese', but he invents the word 'Karyomitose' to describe the overall process of indirect nuclear division and refers to mitotic figures as 'Mitosen'. 'Karyomitose' he subdivides into eight stages, some of which are barely distinguishable as separate events. This over-elaboration was not accepted by Flemming's contemporaries, nor is it to the present day, when pro-, meta-, ana- and telophase, although themselves arbitrary subdivisions, seem to suffice.

In 1880 the second paper[20] in the series appeared in the same journal. Observations on indirect nuclear division are now extended to a wide range of cells, and there is further criticism of Strasburger who had proposed that the mechanism of nuclear division might differ in different types of cell. The third paper[21] appeared in the following year and is eighty-six pages long. It contains a further acerbic polemic against Strasburger who, in the meantime, has modified his position, although it must be repeated that even in the third edition of his book, *Zellbildung und Zelltheilung*, published in 1880, he still argues for transverse division of the chromosomes at the metaphase plate except by chance. To this Flemming replies:

> Secondly, let me make it clear that longitudinal splitting of the threads in the equatorial plane, which Strasburger is obliged to propose in order to accommodate the data, will not serve his purpose: for, as I have described in great detail, the longitudinal splitting of the threads, which I discovered in Salamander, begins at the stage when the threads appear as a skein . . . and continues throughout the stage when these threads have the appearance of a star.*

In this paper Flemming reports his observations on division of the nuclei in the embryo sac of *Lilium croceum* and other plants. He finds that the phenomena he had previously described in the salamander, including longitudinal splitting of the chromosomal threads, applies to plants also. He finds the same thing when he examines fertilization and division of the egg in three different echinoderms, but he is very argumentative about minor disparities between his own observations and those of Hertwig, Fol, Selenka and Schneider, who will be discussed presently. Finally, he turns his attention to mitosis in the corneal epithelium of man and reports that it is no different. In 1882 Flemming produced a book, *Zellsubstanz, Kern und Zelltheilung*[22] (Cell substance, nucleus and cell division) in which he collects together the results of his researches. He naturally emphasizes, and stoutly defends, his principal discovery, the longitudinal splitting of the chromosomes, but he is no more curious in his book than he was in papers about why it is so important that the chromosomes should divide longitudinally and not transversely. Indeed, he sets his face against theorizing of any kind and dismisses contemporary hypotheses in a single paragraph. He is, however, quite clear that nuclei do not arise spontaneously out of amorphous organic matter and, taking a leaf out of Virchow's book, he encapsulates this conviction in the aphorism 'Omnis nucleus e nucleo'.

The principal impetus for the transition from purely descriptive to analytical chromosome cytology was, however, the accumulation of data concerning the fertilization and early development of the egg. In many German texts Bütschli is cited as the originator of this movement, and it does not belittle his work in any way to find that he had a forerunner of whose presence he was totally unaware. The formation of the two pronuclei that make their appearance after fertilization of the egg by the sperm was first described not by Bütschli, but, twenty years earlier, by Nicholas Warneck. Warneck published his work in the *Bulletin of the Imperial Society of Naturalists*, in Moscow, in 1850,[23] but by the 1870s his work was entirely forgotten, if not ignored from the very beginning, until it was resurrected by Hermann Fol. Warneck's work dealt with the development of the fertilized egg in the freshwater gastropods, *Limax* and *Limnoeus*. He noted that there were at first two rounded bodies in the developing egg, but that later there was only one. Bütschli's observations were made on the developing eggs of nematodes, and although in 1873 they were naturally more detailed than those of Warneck, the work was essentially confirmatory.[24] Richard Goldschmidt, in a panegyric on Bütschli written in 1953,[25] perhaps reads more into Bütschli's observations than Bütschli himself intended. For example, Bütschli[26] did not demonstrate that one of the pronuclei in the fertilized egg was derived from the sperm. In fact, he suggested other possibilities:

> After the egg has remained in this condition for some time, I have constantly observed the appearance of a bright vesicle at the pole facing the vagina, and, after a while, a second such vesicle close by. . . . Of course, the origin of these vesicles cannot be seen by direct inspection; they become visible only after they have reached a certain size. One cannot rule out the possibility that the second vesicle is a derivative of the first, although this does seem improbable.*

61. Oskar Hertwig (1849–1922)

Moreover, Bütschli's early papers do not contain any unequivocal statement that might confirm Goldschmidt's claim that Bütschli was the first to show that only one sperm penetrates the egg.

It appears to have been Hermann Fol, a Frenchman who was then a professor in Geneva, who first provided decisive evidence that this was indeed the case.[27] On the other hand, as mentioned earlier,[28] it was Bütschli who demonstrated that the structures seen by Balbiani in mating protozoa, and thought by him to be testicles, were actually involved in the division of the micronucleus. And Bütschli did provide excellent descriptions of all the stages of mitosis in animal cells; but then, so did several other authors at about the same time. Fol's chosen material was the starfish, but he also worked with the eggs of the sea urchin and the arrow worm. His description of a single sperm penetrating the ovum is unambiguous:[29] 'Penetration takes place at some point on the surface of the vitellus. It is my view that normal fertilization of the starfish involves only a single zoosperm for each egg; in the sea urchin this is perfectly clear.'* Fol actually observed the transfer of the intact sperm nucleus into the egg cytoplasm and the development of this nucleus into the male pronucleus: 'The point at which penetration occurs becomes the centre of a star or male aster; in the middle of the aster there is an accumulation of material that forms the male pronucleus, and this then fuses with the female pronucleus in a manner that is altogether similar to what one observes in the sea urchin.'*

Some German texts stress that Fol was anticipated by Oskar Hertwig, at that time a *Privatdozent* and later on an *Extraordinarius* at the University of Jena. In a series of papers, the first of which was published in 1875,[30] Hertwig

62. Edouard van Beneden (1846–1910)

proposed that, in the fertilized egg of the sea urchin *Toxopneustes lividus*, one of the pronuclei arose from the nucleus of the sperm, but his proposal was based on inference, not on direct observation. He believed that the nucleus of the zygote was produced by direct fusion of the two pronuclei, and this view was supported by Fol, Selenka and Flemming. Actually, it appears that Hertwig was, in turn, anticipated by Auerbach. Writing from Breslau in 1874,[31] Auerbach reported that the two pronuclei fused together to form the nucleus of the zygote. This amalgamation of the nuclear material derived from the male with that derived from the female Auerbach considered to be the essential feature of sexual reproduction. And so it was, for it meant that in the nuclei of all the cellular progeny of the fertilized egg there were both paternal and maternal contributions.

Hertwig also believed, under the influence of Bütschli, that the chromosomes were swellings of the spindle fibres. Although Hertwig's views on the formation of the male pronucleus and the zygote nucleus were derived from his observations on sea urchin eggs, his work extended to heteropods, leeches and frogs, and he evidently considered that, at least for animal eggs, his findings were quite generally applicable.[32] Selenka's observations were made on a different species of sea urchin, *Toxopneustes variegatus*, and were, in principle, similar to those of Hertwig and Fol. They are collected in a book which he published in 1878,[33] although an earlier, undated, contribution is cited by van Beneden.[34] It is, however, clear from Selenka's work that the apparent fusion of the two pronuclei into the zygote nucleus had stimulated investigators to ask questions about the role of the nucleus in general and, in particular, to explore what the paternal and maternal contributions to it might be. It was perhaps a question of this kind that induced Selenka to count the chromosomes in *Toxopneustes variegatus*, but regrettably he found, no doubt for technical reasons, that the number varied.

Van Beneden's classic paper is in effect a monograph and is 375 pages long.

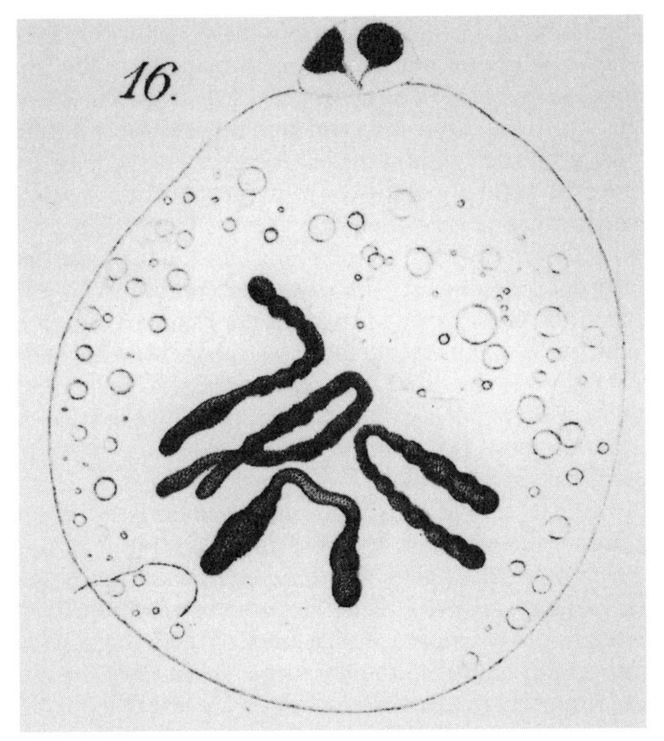

63. Van Beneden's illustration of the four chromosomes ('anses chromatiques') of *Ascaris maglocephala*, two paternal and two maternal

It appeared in the *Archives de Biologie* in 1883.[35] Van Beneden was born in Louvain, held a chair of zoology at Leiden, and died in Liège. Belgian academic life was at that time entirely francophone, and van Beneden naturally wrote in French. The factor that enabled him to clarify so many of the points of uncertainty in the study of fertilization was his choice of experimental object, *Ascaris megalocephala*. The eggs of this parasitic nematode of the horse are large and transparent, but, more importantly, there is synchrony of the events following fertilization as one passes down the uterus, so that large numbers of fertilized eggs at any desired stage of development are easily obtainable. What is more, this species of *Ascaris* has only four chromosomes, and some species even have only two, so that the confusion resulting from large numbers of overlapping chromosomes does not arise. Van Beneden categorically denied that the two pronuclei of the fertilized egg fused together ('les deux pronucléus ne se confondent jamais').

His paper begins with an exhaustive review of the work of previous investigators, including those who claimed to have observed direct fusion of the pronuclei, and then sets out in great detail precisely what he sees in the eggs of *Ascaris*. His conclusions are as follows:

> (1) The formation of the male pronucleus involves not only the chromatic nucleus of the zoosperm but also the achromatic layer that surrounds it (the perinuclear layer). (2) The germinal vesicle contributes to the female pronucleus not only chromatic elements, but also an achromatic body. (3) The two pronuclei, without fusing together, can acquire, by virtue of their progressive maturation, the constitution of ordinary nuclei. (4) In the horse threadworm a single nucleus is not formed at the expense of the two pronuclei; a 'Furchungskern' (cleavage nucleus), in the sense that Hertwig uses the word, does not exist. The essence of fertilization does not therefore reside in the conjugation of two nuclear elements, but in the formation of these two elements within the female gonocyte. One of these nuclei comes from the egg, the other from the zoosperm. The nuclear elements ejected in the form of polar globules are replaced by the male pronucleus, and as soon as these two half-nuclei, one male, the other female, are formed, fertilization is complete. (5) Following a series of transformations which the nuclear scaffold of each pronucleus undergoes, transformations that are identical, moreover, to those that take place in any dividing nucleus, each pronucleus gives rise to two chromatic loops. (6) The four chromatic loops are involved in the formation of the chromatic star (metaphase plate); but they remain distinct. Each of them then divides longitudinally into a twin pair of secondary loops. (7) Each of the nuclei of the first two blastomeres receives one half of each primary loop, that is to say, there are four secondary loops, two male and two female.
>
> Thus there is no fusion between the male chromatin and the female chromatin at any stage of division. . . .
>
> The elements of male origin and those of female origin are never fused together in a cleavage nucleus, and perhaps they remain distinct in all the nuclei derived from them.*

There is hardly a word of this that we would now wish to change, and it is amply illustrated by drawings of the greatest clarity. But the essential feature of van Beneden's study is that it is concerned with biological principles, not simply anatomical descriptions. The secondary loops are referred to as twins, and the fact that their morphology, as shown in the illustrations, is identical to that of the primary loop from which they are derived implies that there is an identity of function underlying the identity of structure. From the observation that each of the two primary blastomeres receives one half of each primary loop, it is clear that van Beneden is aware that the nuclei of the two primary blastomeres receive identical chromatic loops. And this is confirmed by his surmise that these loops are maintained as distinct and stable entities in all the progeny of the primary blastomeres.

In later work van Beneden extended his studies to mammals, particularly rabbits and bats. In these he found nothing that contradicted his observa-

64. Carl Rabl (1853–1917)

tions on *Ascaris*. Van Beneden and Neyt,[36] in 1887, also unravelled much of the mystery surrounding the formation of the centrosome (a small specialised region in the cytoplasm essential for accurate separation of the chromosomes). They regarded this organelle, with its attraction sphere and central corpuscle, as having a permanent life of its own. It was to be found not only in the blastomeres, but also in all their progeny. Each attraction sphere apparently arose from a pre-existing attraction sphere, and each corpuscle from a pre-existing corpuscle. What is more, the division of this organelle was shown to precede that of the nucleus. So impressed were van Beneden and Neyt with the independence of the centrosome that they even gave it as their opinion that it was no less important than the nucleus. In a postscript to his 1883 paper[37] van Beneden mentions the work of Schneider which appeared in book form in the same year[38] and apparently reached van Beneden after his own paper had gone to the printer. Schneider also worked with *Ascaris megalocephala* and, as van Beneden admits, also saw karyokinesis in the egg pronucleus; but Schneider denied the existence of a sperm pronucleus and considered the two pronuclei of the fertilized egg to be the division products of the egg nucleus. Van Beneden is merciless in his criticism. The polemical nature of this postscript is reminiscent of Schleiden's reference to Raspail: van Beneden also urges the reader to judge for himself the worthlessness of Schneider's book. Van Beneden's observations on *Ascaris* eggs were confirmed in the following year by Heuser,[39] who found much the same thing in plants.

Carl Rabl, an Austrian who was Virchow's son-in-law and whose work was much admired by Boveri, was no less meticulous in his observations than van Beneden, but perhaps more acutely conscious of their genetic consequences, or at least more overtly so. Rabl's paper was published in 1885[40] and deals mainly with cell division in the epidermis of the mouth and gills in the larva of *Salamandra maculata*. This, although good material for cytological studies, is less favourable than *Ascaris megalocephala*, and the strictly anatomical findings described by Rabl are therefore less detailed than those reported in van Beneden's paper which appeared two years earlier. Nonetheless, Rabl was able to show that the chromosomes in prophase were not connected to each other in a skein (Knäuel), as Flemming had suggested, but were separate entities from the beginning. The chromosomes in prophase were actually numbered and this same number was found to be present in metaphase. Finally, Rabl demonstrated that the chromosome number was constant for each species, thus disproving the variable numbers found in *Toxopneustes* by Selenka.[41] According to Mayr,[42] Rabl was the first 'to formulate clearly' what Boveri later called the individuality and the continuity of chromosomes. But this seems to me to attribute more to what Rabl inferred from constant chromosome numbers than the findings themselves perhaps warrant. Van Beneden challenged Rabl's priority in this matter, and there is no doubt that his own claim is strongly supported by his observation that both the number and the morphology of the chromosomes were constant in *Ascaris*. Moreover, van Beneden showed that this constancy was maintained in the blastomeres, and he surmised, in print, that it was also maintained in all the progeny of the blastomeres. But Rabl's work was more widely appreciated by German writers, partly for the usual reasons, and partly because Rabl's observations, having been made mainly on vertebrate cells, were thought to have general validity.

In microanatomical studies of great precision, nomenclature is obviously critical. It has already been noted that the term 'karyokinesis' was introduced by Schleicher[43] and the term 'mitosis' by Flemming.[44] 'Prophase', 'metaphase' and 'anaphase' were coined by Strasburger in a paper that appeared in 1884.[45] It is in this paper that Strasburger relents. He not only admits that the chromosomes divide by longitudinal splitting, he even brings himself to say that Flemming's contribution to the complicated subject of nuclear division has been outstanding. He defines prophase in the following terms: 'the introductory phases of nuclear division, which I propose to call "prophases", begin with the formation of the skein of thread.'* Strasburger's definition of metaphase concedes the longitudinal division of the chromosomes: 'In contrast to the prophases, which extend to the longitudinal splitting of the segments in the nuclear plate, those phases that begin with the separation of the daughter segments and eventually result in their complete segregation and rearrangement I shall subsume under the name "metaphases".'* And anaphase is defined as follows: 'The phases that extend from the complete separation of the daughter segments . . . to the reconstitution of the daughter nuclei can be collectively called anaphases of nuclear division.'*

The word chromosome, to replace the previous miscellany of 'Stäbchen', 'Schleifen', 'Fäden' (and 'bâtonnets') was proposed by Waldeyer (H. W. G.

Waldeyer-Harz), but not until 1888.[46] Waldeyer, after holding chairs at several German universities, was by then professor of anatomy in Berlin, where he died. Telophase did not arrive until 1894. In a massive paper,[47] Martin Heidenhain, a former *Assistent* of Köllikers, uses the words 'Telophasen' and 'Telokinesis' in the heading of his Chapter 4. In this chapter he gives the following definition: 'I describe by the name telokinesis certain movements of the nucleus and the microcentre that take place towards the end of mitosis. These movements run their course in a manner that is at the very least absolutely characteristic in so far as it always produces an outcome that is completely clearcut.'* Heidenhain then goes on to give a detailed justification of his use of the new name.

The term 'interphase' made its appearance in the year before the outbreak of the First World War. Lundegårdh[48] invented it to make a distinction between the resting cell that does not proceed to mitosis, and the apparently inactive phase that a cell undergoes as it proceeds from one mitosis to the next:

> When the so-called resting nucleus is analysed in detail it transpires that there is a difference in the configuration of the caryotin between cells that have just entered the resting phase and those that have been in it for a long time. It therefore seemed appropriate to give a special name to the interval between two successive cell divisions. I call it an interphase and speak of a nucleus being in a typical resting phase only when interphase has gone on for a long time.*

The work described in this chapter is essentially morphological in character. It is concerned, for the most part, with the structure of chromosomes, their number, disposition and replication. Both van Beneden and Rabl clearly had an insight into the biological significance of the phenomena they described, but their papers dealt only peripherally with the question of function. Function did not become the centre of attention until the main morphological features of nuclear division were clear. How the function of chromosomes was initially analysed and how the relationship between chromosomal behaviour and the transmission of hereditary characteristics was established form the subject of the following chapter.

CHAPTER 17

The Cellular Determinants of Heredity

It is not the object of this book to present an abbreviated history of genetics; however, no account of the cell doctrine can omit altogether a discussion of the cellular determinants of heredity and how they were discovered. It appears that one of the first to propose that the nucleus contained the hereditary material of the cell was Haeckel, who for more than forty years was professor of zoology at Jena. Haeckel's writings are so voluminous and so permeated with theories that have subsequently been discredited, that it is difficult to say whether his views on the nucleus were based on empirical evidence or whether they were merely an inspired conjecture among many that were less inspired. In any case, in his *Allgemeine Anatomie* of 1866[1] he contrasts his 'cytodes' that do not contain a nucleus, and may or may not be bounded by a membrane, with real cells that do contain a nucleus. 'We must distinguish between the elementary organisms that we have defined as cytodes and real cells which contain a nucleus.'* And again: 'The cytodes or anucleate lumps of plasma fall into two groups as do real cells which contain a nucleus.'* Haeckel obviously regarded the nucleus as an indispensable component of a real cell and actually proposed that it was responsible for the transmission of hereditary characters; the cytoplasm, he suggested, was concerned with the accommodation of the cell to its environment. This inspired guess was not generally accepted and almost two decades were to elapse before it was taken seriously, if the brief statement that Haeckel buried in two large volumes was remembered at all.

In the early 1880s several investigators began to concern themselves with the transmission of hereditary characteristics at the level of the cell. The plant-breeding experiments of Kölreuter, which were done in the 1760s,[2] had shown that, at least in *Nicotiana*, the heritable contributions of egg and sperm were equal. But the mass of egg cytoplasm was vastly greater than that of the sperm. This self-evident observation induced Nägeli to formulate an imaginative theory of inheritance that attracted a great deal of attention at the time, but soon collapsed. From the disparity between the 'nutritive plasm' of egg and sperm, Nägeli concluded[3] that the whole of the cytoplasm could not be involved in the transmission of hereditary characteristics. These, he postulated, were carried by a small part of the cytoplasm that was stable and not sensitive to normal environmental influences. To this component he gave the

65. Ernst Haeckel (1834–1919)

name 'idioplasm'. To conform to the known facts, the 'idioplasm' had, of course, to be present in the sperm and the ovum in equal amounts. Nägeli then proceeded to describe in detail the physical properties of the idioplasm, a description that was totally unattached to reality. The idioplasm, which passed from cell to cell, was supposed to consist of long strands with a micellar structure that was everywhere identical in cross-section. Growth was achieved by elongation of the strands without any change in their underlying structure. Each strand nonetheless had specific properties and, in virtue of these, controlled the differentiation of cells and tissues. It can be said in favour of this fantasy that it shows Nägeli to have had a clear idea of the distinction between the determinant of an inherited trait and the trait itself. But it remains surprising, given the information that was already available, that Nägeli nowhere implicates the nucleus as a possible vector for his idioplasm.

In the year preceding the appearance of Nägeli's book, Wilhelm Roux, then a *Privatdozent* at the University of Breslau, published a theoretical article[4] that gave a definitive explanation for the fact that nuclear division was 'indirect' and involved the longitudinal splitting of the chromosomes. Roux argued that 'indirect' nuclear division ensured the equal partitioning not only of the quantity of nuclear material, but also of its quality. While 'direct' nuclear scission might achieve accurate partitioning of the nucleus into two equal halves, it would achieve equal division of the properties of the nucleus only if the nucleus was homogeneous, which countless observations indicated was not the case. Roux proposed that each of the chromosomes carried a different panoply of heritable determinants and that longitudinal splitting was therefore essential to ensure that each of the daughter cells received the

66. Wilhelm Roux (1850–1924)

same inheritance. The criticism is sometimes made that Roux undermined his own position by suggesting that the polarity of the embryo was determined by unequal division of the nuclei of the blastomeres; but when one considers that it was not until the second half of the twentieth century that plausible models to account for this polarity became available, it is not surprising that Roux failed to find in the exact partitioning of the heritable qualities of the nucleus an answer to the problem of differentiation. Roux was, after all, an embryologist, and it is something of a compliment to him that he was fully aware that the central problem of embryology was not solved by the known facts of chromosome mechanics. In any case, Roux did not for a moment entertain the possibility that the cellular determinants of heredity might be found anywhere other than in the nucleus.

In the following year, Strasburger[5] provided well nigh decisive experimental evidence in support of this view. Working with the orchid, *Orchis latifolia*, Strasburger showed that when the pollen tube passes down the pistil into the embryo sac, the nucleus is forced out of the end of the tube into the sac, while the cytoplasm is excluded. It could be objected that a trace of cytoplasm might have accompanied the denuded nucleus, but this objection could also have been made to the celebrated experiment reported by A. D. Hershey and M. Chase in 1952.[6] Hershey and Chase showed that when a susceptible bacterium was infected with a bacteriophage, it was the phage nucleic acid, labelled with radioactive phosporus, that was injected into the bacterium, while the phage protein, labelled with radioactive sulphur, was excluded; and this experiment was at once regarded as decisive. Be that as it may, it was difficult, after Strasburger's observations, to ignore the possibility that the nucleus might be involved in the transmission of hereditary traits.

67. August Weismann (1834–1914)

Hertwig made this clear in two papers published in 1884[7] and 1885.[8] In these, Hertwig proposed a new theory of inheritance and pointed out that the specifications required by Nägeli's idioplasm would be met by the compound nucleus formed by amalgamation of the male and female pronuclei. Hertwig even went so far as to suggest that the nuclear material that acted as the vector of inherited characteristics was 'nuclein', a viscous material rich in phosphorus that Friedrich Miescher[9] had extracted from nuclei in 1869. This idea, after a brief period during which it attracted some sporadic interest, lay fallow until it was resurrected by Oswald Avery and his colleagues[10] in the 1940s. Their work, although painstaking in the extreme, was also criticized on the grounds that the samples of nucleic acid that they prepared might have been contaminated by traces of protein.

The role of August Weismann in the history of genetics is too well documented to require an abridgement here. But it is of interest that his two famous works[11] on the germ plasm appeared in the year before and the year after the publication of Nägeli's book on the idioplasm. Weismann made a fundamental distinction between the germ line, which he believed was spatially segregated, and the somatic cell line which gave rise to the rest of the organism. He was convinced that the cell nucleus was the repository of the transmissible hereditary material, and he even postulated that this was disposed in a linear manner along the chromosomal loops. He argued that, at fertilization, a new combination of chromosomes, and hence a new combination of hereditary determinants, must be formed. This argument yielded a satisfactory explanation for the reduction division of meiosis which, in

68. Theodor Boveri (1862–1915)

cytological terms, appears to have been first described in detail by van Beneden. One can only conjecture why it was that Nägeli, apparently deliberately, eschewed the nucleus. It is possible that he was still to some extent under the influence of his own earlier observations on apparently anucleate cells and cells in which the nucleus was merely transient.

The definitive demonstration that the chromosomes were the vectors of inherited traits came from the work of Theodor Boveri who spent most of his working life at the University of Würzburg. His experiments marked a major transition in methodology. Prior to Boveri, analysis of cellular components was based either on theory, as in the case of Weisman, or on minute microscopic examination of selected cells and tissues, as in the case of Flemming. With Boveri the science of cell biology moved from observation and deduction to manipulative experimentation. His work on the function of chromosomes began with a re-examination of the observations of van Beneden, to whom he gives ample credit.[12] Like van Beneden, Boveri chose the eggs of *Ascaris megalocephala* as his experimental material, and the conclusions that he reached were essentially an extension of the views put forward by van Beneden and Rabl. However, Boveri converted into a principle what had previously been a surmise or an interpretation. This he did by extending his observations to *Ascaris univalens*, which has only two chromosomes whose morphology can readily be resolved. With this species Boveri was able to show convincingly that the number, disposition and morphology of the chromosomes seen during karyokinesis of the blastomeres was preserved. He insisted, moreover, that this observed stability was quite general and he accorded it the status of a principle which he entitled 'The Continuity of Chromosomes'.

In essence, continuity of chromosomes meant, in Boveri's own words, that the chromosomes were 'selbstständige Individuen, die diese Selbstständigkeit auch im ruhenden Kern bewahren' (independent entities that preserve their independence even in the resting nucleus). What comes out of the nucleus, he argued, was what goes into it. This was, of course, Rabl's thesis, but it now rested on much firmer evidence. In 1888,[13] still working with *Ascaris*, Boveri also confirmed the independence of the centrosome and delineated its life-cycle. Van Beneden and Neyt[14] had come to very similar conclusions a year earlier, but Boveri was apparently unaware of this.

It was, however, Boveri's second principle, 'The Individuality of Chromosomes',[15] that demonstrated his remarkable experimental skill. He now turned to sea urchin eggs, the material introduced by Hertwig. These were chosen because it was possible by relatively simple procedures to produce blastomeres with different and incomplete chromosome sets. If the sea urchin eggs were exposed to very high concentrations of sperm, the introduction of two sperms into a single egg was commonly achieved, with the result that tripolar and tetrapolar mitoses were subsequently formed. Tetrasomy could also be induced by inhibiting the first cleavage division of the egg and thus generating a cell with four centrosomes. Boveri already knew that the chromosomes retained their continuity from one cell generation to the next; that the developing egg received chromosome complements from both egg nucleus and sperm nucleus; and that each parental chromosome set was itself enough to permit normal development of the egg. By studying the fate of blastomeres with experimentally induced abnormalities in chromosome constitution,[16] he was therefore able to test the proposition put forward by Roux, that each chromosome carries a different genetic load. The results that Boveri obtained showed decisively that Roux was right. 'Verschiedenwertigkeit der Chromosomen als Erbträger' (The different capacities of individual chromosomes to transmit heritable traits) was thus established by direct experimental intervention.[17]

These findings were not, however, universally accepted. They were challenged by Hertwig and notably by Fick.[18] Indeed, even as late as 1909[19] Boveri was still defending his position against Fick in a manner that we might regard as polemical. Boveri's own summary of his final position would not be out of place in a modern textbook:

> At fertilization, these two 'haploid' nuclei are added together to make a 'diploid' nucleus that now contains 2a, 2b and so on; and, by the splitting of each chromosome and the regulated karyokinetic separation of the daughter chromosomes, this double series is inherited by both of the primary blastomeres. In the resulting resting nuclei the individual chromosomes are apparently destroyed. But we have the strongest of indications that, in the stroma of the resting nucleus, every one of the chromosomes

> that enters the nucleus survives as a well-defined region; and as the cell prepares for its next division this region again gives rise to the same chromosome (*Theory of the Individuality of the Chromosomes*). In this way the two sets of chromosomes brought together at fertilization are inherited by all the cells of the new individual. It is only in the germinal cells that the so-called reduction division converts the double series into a single one. Out of the diploid state, the haploid is once again generated.*

Boveri was not slow to see the correspondence between the individuality of chromosomes and the segregation of factors described by Gregor Mendel in his *Versuche über Pflanzen-Hybriden* which appeared in printed form in 1866. As is well known, Mendel's paper, long forgotten, was resurrected by de Vries, Correns and Tschermak in 1900. Boveri appears to have first referred to this parallelism in 1902[20] and again in 1903,[21] and to have discussed it in more detail in 1904.[22] He could not avoid noticing that the stability and individuality of the inherited traits described by Mendel could be readily accommodated by the chromosome mechanics that he had done so much to clarify. He did not again doubt that it was the behaviour of the chromosomes that determined the rules of inheritance Mendel had discovered.

The turn of the century saw the beginnings of an American contribution to the cell doctrine. In 1901 Thomas H. Montgomery, who was then associated with the University of Pennsylvania and the marine biological laboratory at Woods Hole, published his studies[23] on the spermatogonial chromosomes of forty-two species of hemiptera. He found that the chromosomes differed greatly in size, particularly the largest ones and the smallest, presumably the sex chromosomes. He also found that the paternal chromosomes always paired at meiosis with the maternal ones, commenting that this must obviously also be the case in *Ascaris univalens* which has only two chromosomes. There was thus a strong inference that the chromosomes differed in their qualities, as well as in their size. Montgomery's paper does not give evidence of any attempt to confirm this inference by experiment, and it is diluted by a good deal of tenuously based speculation.

C. E. McClung's paper,[24] which appeared in 1902, was entitled 'The accessory chromosome–sex determinant'. Working mainly with grasshoppers, McClung, then an instructor of zoology at the University of Kansas and later the head of the department there, found that two kinds of sperm were formed, only one of which contained an accessory chromosome. Since this accessory chromosome was present in one half of the sperm cell population, McClung drew the obvious conclusion that it must be involved in determining the sex of the resulting individual. This may well have been one of the first attempts to localize a specific function to a particular chromosome, but it was far from clear that the functional specificity that McClung had found for the accessory chromosome was generally applicable to the other chromosomes. It was Sutton who, also in 1902, showed that this was in all probability the case.[25] Sutton was McClung's first graduate student, moving in 1901 to E. B. Wilson's laboratory at Columbia, New York. Sutton found that in the lubber grasshopper *Brachystola magna* each of the eleven pairs of chromosomes was morphologically different from the others, including, of course, the small accessory chromosome. It therefore seemed very likely 'that the constant morphological differences between the ordinary chromosomes are the visible expression of physiological or qualitative differences'. At the conclusion of his paper, Sutton suggests that 'the association of paternal and maternal chromosomes in pairs and their subsequent separation during the reducing division as indicated above may constitute the physical basis of the Mendelian law of heredity'.

This point is further elaborated in a review that Sutton published in 1903.[26] Since E. B. Wilson, in whose laboratory the work had been done, sought to propagate the name 'Sutton–Boveri' theory for these ideas, it is perhaps worth mentioning that Sutton was aware of Boveri's work when he wrote his 1902 paper, and so was Wilson who had actually worked with Boveri. In fact, Sutton expressly says so, in a passage that does him credit: 'The evidence advanced in the case of the ordinary chromosomes is obviously more in the nature of a suggestion than of proof, but it is offered in this connection as a morphological complement to the beautiful experimental researches of Boveri already referred to.' And that sums up the difference between Montgomery, McClung and Sutton on the one hand and Boveri on the other. The first three all make inferences from morphological observations only, whereas Boveri seeks to settle the matter by experiment.

In the year that saw the outbreak of the First World War, Boveri wrote his extraordinarily prophetic monograph on the origin of malignant tumours,[27] which advances a theory that is not too far removed from our present ideas on this subject. He died a year later. By that time, to argue that hereditary traits were transmitted by anything other than the chromosomes had become an eccentricity, and the question was no longer whether the chromosomes functioned in this way but how they did so. This new phase in the study of chromosomal function was initiated by Thomas Hunt Morgan. Morgan's first papers on the genetics of the fruit-fly *Drosophila melanogaster* appeared in 1910[28] while Morgan was still a member of E. B. Wilson's institute at Columbia, and it was not long before this insect, which bred very rapidly, became the centrepiece of genetical analysis. Morgan, who subsequently moved to the California Institute of Technology, gathered around him a number of gifted collaborators and the cumulative contributions of this group eventually became canonical.[29]

A new dimension to the work was opened up by Theophilus Painter who had also worked with Boveri. In 1933[30] he reported the discovery of the giant chromosomes in the salivary glands of *Drosophila* and realized that they offered an unparalleled opportunity for the investigation of chromosome structure.[31] It at once became possible to correlate mutations in the adult fly with visible alterations at specific loci in the giant chromosomes. In August 1914 work in the main biological laboratories of Europe was suspended, and the initiative in the genetical analysis of chromosome function passed to the laboratories of the United States where European scholars, fleeing the horrors of National Socialism, made an immense contribution.

By the early part of the twentieth century the main outlines of the cell doctrine had been established. It was agreed that both animal and plant tissues were essentially composed of cells; that the cells multiplied by binary fission; that they consisted of protoplasm which was bounded by a membrane and contained a nucleus; that the nucleus was the repository of chromosomes which became visible and split along their length when the cell divided; that the chromosomes were the vectors of heritable characters and that each chromosome had a specific morphology and a specific function. These facts, amalgamated with the theory of natural selection, form the bedrock of all modern life sciences. Whatever the origin of life might have been, the universal solution to the problems confronting its further evolution was the progressive assembly of the cell. That is the greatest contribution to human knowledge that biology has to offer.

Notes

CHAPTER 1

1. The later editions of Diels–Krantz remain the standard work on the surviving fragments of the pre-Socratics.
2. J. Barnes, *The Presocratic Philosophers*. Routledge & Kegan Paul, London (1979). Barnes presents a detailed analysis of the philosophy of atomism, but comes down heavily on the side of Aristotle.
3. An informative summary of the works of Epicurus, including bibliographical sources, is to be found in *The Oxford Classical Dictionary* (2nd edn, N. G. L. Hammond and H. H. Scullard, eds). Clarendon Press, Oxford (1970).
4. See H. J. Cook, 'The New Philosophy in the Low Countries' in *The Scientific Revolution in National Context* (R. Porter and M. Teich, eds). Cambridge University Press, Cambridge (1992), p. 115.
5. See L. W. B. Brockliss, 'The Scientific Revolution in France', ibid., p. 55.
6. *Marcello Malpighi, Opere scelte a cura di Luigi Belloni* (Selected works). Unione tipografico, Editrice Torinese (1967).
7. Each of these authors is discussed at a later stage.
8. Hooke's drawings of a section through cork and of the underside of a nettle leaf are reproduced in C. Singer, *A Short History of Biology*. Clarendon Press, Oxford (1931).
9. Dr Lionel Clowes of the Department of Plant Sciences in the University of Oxford has confirmed this in living specimens.
10. J. Sachs, *Geschichte der Botanik von 16 Jahrhundert bis 1860.* R. Oldenbourg, Munich (1875).
11. A. G. Morton, *History of Botanical Sciences.* Academic Press, London and New York (1981).
12. There is a discussion of the contributions of these early Italian microscopists in Belloni (note 6).
13. A. Pollender, *Wem gebührt die Priorität in der Anatomie der Pflanzen dem Grew oder dem Malpighi?* Georgi, Bonn (1868).
14. See note 6.
15 *The Correspondence of Marcello Malpighi.* H. B. Adelmann (ed.). Cornell University Press, Ithaca and London (1975).
16. Leeuwenhoek's letters were written in Dutch and shortened versions of them, in English, were published by the Royal Society of London to whom they were sent. The full Dutch text is given in *The Collected Letters of Antoni van Leeuwenhoek*, edited, illustrated and annotated by a Committee of Dutch scientists. Swets & Zeitlinger Ltd, Amsterdam (1948).
17. J. R. Baker, *Quarterly Journal of Microscopical Science* **89** (1948): 103.

CHAPTER 2

1. Epistles I, I, 14.
2. Quoted by W. Clark, *The Scientific Revolution in the German Nation*, in *The Scientific Revolution in National Context* (ed. R. Porter and M. Teich) p. 90. Cambridge University Press (1992).
3. See Chapter 1, note 4.
4. J. Swammerdam, *Biblia naturae* (ed.

H. Boerhaave) Vol. 1, p. 69. Severinum, Vander & Vander, Leiden (1737–80).

5. See Chapter 1, note 16.
6. W. Turner, *Nature, Lond.* **43** (1890): 10.
7. K. F. Wolff, *Theoria generationis.* Halle (1759). The quotations are from the second edition: *Theoria generationis, editio nova, aucta et emendata.* Christ. Hendel, Halae ad Salam (1774).
8. F. K. Studnička, *Acta Soc. Scient. Natural. Moravicae* **4** (1927): fasc: 4: 1.
9. F. Fontana, *Traité sur le venin de la vipère . . .* Florence (1781).
10. J. Skinner, *Treatise on the Venom of the Viper . . .* John Murray, London (1787).
11. A. Monro, *Observations on the Structure and Functions of the Nervous System.* Edinburgh and London (1783).
12. C. F. Heusinger, *System der Histologie.* Bärecke, Eisenach (1822).

CHAPTER 3

1. See Chapter 1, note 10.
2. M. J. Schleiden, *Arch. Anat. Physiol. Wiss. Med.* (1838) p 137.
3. See Chapter 2, note 8.
4. F. K. Studnička, Anatom. Anzeiger **73** (1932): 390.
5. P. Broca, *Traité des tumeurs.* Asselin, Paris (1866).
6. M. Klein, *Histoire des origines de la théorie cellulaire.* Hermann, Paris (1936).
7. F. Duchesneau, *Genèse de la théorie cellulaire.* Bellarmin, Montréal, Vrin, Paris (1987).
8. J. Schiller and T. Schiller, *Henri Dutrochet (du Trochet) 1776–1847.* Blanchard, Paris (1975).
9. X. Bichat, *Anatomie générale appliqué à la physiologie et à la médecine.* Brosson, Gabon, Paris (1801).
10. H. Milne-Edwards, *Mémoire sur la structure élémentaire des principaux tissus organiques des animaux.* Lejeune, Paris (1823).
11. H. Milne-Edwards. *Répertoire générale d'anatomie et de physiologie.* Vol. 3. Paris (1827).
12. H. Dutrochet, *Recherches anatomiques et physiologiques sur la structure intime des animeaux et des végétaux, et sur leur motilité.* Baillière, Paris (1824).
13. See note 4.
14. F. Arnold, *Lehrbuch der Physiologie des Menschen.* Zürich (1836).
15. H. E. Weber, *Allgemeine Anatomie des menschlichen Körpers.* Vol. 1 of F. Hildebrandt's *Handbuch der Anatomie des Menschen.* 4th edn., Verlag der Schulbuchhandlung, Braunschweig (1830).
16. A. R. Rich, *Bull. Johns Hopkins Hosp.* **39** (1926): 330.
17. T. Schwann, *Mikroskopische Untersuchungen über die Uebereinstimmung in der Struktur und dem Wachstum der Thiere und Pflanzen.* Sander'schen Buchhandlung, Berlin (1839).
18. See note 8.
19. H. Dutrochet, *Mémoires pour servir à l'histoire anatomique et physiologique des végétaux et des animaux.* Ballière, Paris (1837).
20. J. J. P. Moldenhawer, *Beiträge zur Anatomie der Pflanzen.* Wäser, Kiel (1812).
21. See note 12.
22. See note 19.
23. See note 4.
24. See Chapter 2, note 12.
25. E. Du Bois-Reymond, *Untersuchungen über thierische Elektricität* Reimer, Berlin (1848–84).
26. F.-V. Raspail, *Nouveau système de chimie organique, fondé sur des méthodes nouvelles d'observation.* Ballière, Paris (1833).
27. Ibid.
28. D. B. Weiner. *François-Vincent Raspail, 1794–1878. Scientist and Reformer.* Columbia University Press, New York (1968).
29. F.-V. Raspail, *Histoire naturelle de la santé et de la maladie chez les végétaux et les animaux en général, et en particulier chez l'homme.* Levasseur, Paris (1843).
30. F.-V. Raspail, *Annales des sciences naturelles* **6** (1825): 224 and 384.
31. See note 26.
32. See note 30.

33. L. Oken, *Die Zeugung*. Goebhardt, Bamberg and Würzburg (1805).
34. F. Leydig, *Lehrbuch der Histologie des Menschen und der Thiere*. Meidinger Sohn, Frankfurt (1857).
35. F.-V. Raspail, *Mém. soc. hist. natur. de Paris* **3** (1827): 305.
36. See note 26.
37. F.-V. Raspail, *Nouveau système de physiologie végétale et de botanique*. Baillière, Paris (1837).
38. See note 33.
39. See note 6.
40. P.-J.-F. Turpin, *Mémoires du muséum d'histoire naturelle*. **14** (1827): 15.
41. Ibid; **12** (1826): 161.
42. See note 12.
43. P.-J.-F. Turpin, *Mém. muséum hist. naturelle* **16** (1828): 57.
44. Ibid., **16** (1828): 295.
45. Ibid., **18** (1829): 161.

Chapter 4

1. C. F. Brisseau-Mirbel, *Histoire naturelle des plantes* **8**: 1, 57. (L'an 10=1802.)
2. C. F. Brisseau-Mirbel, *Exposition et défense de ma théorie de l'organisation végétale*. Van Cleef, The Hague (1808).
3. C. F. Brisseau-Mirbel, *Exposition de la théorie de l'organisation végétale*. 2nd edn, Paris (1809).
4. G. R. Treviranus, *Biologie oder Philosophie der lebenden Natur für Naturforscher und Aerzte*, Vol. 3. Röwer, Göttingen (1805).
5. G. R. Treviranus, *Vom inwendigen Bau der Gewächse und von der Saftbewegung in demselben*. Dietrich, Göttingen (1806).
6. See Chapter 3, note 20.
7. A. Hughes, *A History of Cytology*. Abelard-Schuman, London and New York (1959).
8. See Chapter 1, note 10.
9. See note 2.
10 See note 5.
11. D. H. F. Link. *Grundlehren der Anatomie und Physiologie der Pflanzen*. Danckwerts, Göttingen (1807).
12. K. A. Rudolphi, *Anatomie der Pflanzen*. Myliussischen Buchhandlung, Berlin (1807).
13. J. F. Bernhardi, *Beobachtungen über Pflanzengefässe und eine neue Art derselben*. Erfurt (1805).
14. L. C. Treviranus, *Beyträge zur Pflanzenphysiologie*. Dietrich, Göttingen (1811).
15. Ibid.
16. Ibid.
17. G. R. and L. C. Treviranus, *Vermischte Schriften* von G. R. and L. C. Treviranus (1817).
18. L. C. Treviranus, *Annales des sciences naturelles* **10** (1827): 22.
19. B. Corti, *Osservazioni microscopiche sulla Tremella e sulla circolazione di fluido in una pianta aquajuola*. Lucca (1774).
20 See note 11.
21. C. F. Brisseau-Mirbel, *Elémens de physiologie végétale et de botanique*. Magimel, Paris (1815).
22. See note 2.
23. See note 11.
24. Ibid.
25. C. F. Brisseau-Mirbel, *Mém. acad. roy. sci. Institut de France* **13** (1835): 337.
26. See Chapter 1, note 10.
27. K. Sprengel, *Anleitung zur Kenntnis der Gewächse*, 1st edn (1802). Quotations are from 2nd edn, Kümmel, Halle (1817).
28 Ibid.
29. D. G. Kieser, *Mémoire sur l'organisation des plantes*. Beets, Harlem (1814).
30. D. G. Kieser, *Elemente der Phytotomie*. Cröcker, Jena (1815).

Chapter 5

1. See Chapter 3, note 15.
2. F. J. F. Meyen, *Phytotomie*. Haude und Spener. Berlin (1830).
3. See Chapter 3, note 15.
4. J. L. Prévost and J. A. Dumas, *Biblothèque universelle des sciences belles lettres et arts*. **17** (Sciences et Arts), (1821): 215.
5. F. J. F. Meyen, *Neues System der Pflanzen-Physiologie*. Haude und Spener, Berlin 1837).
6. See Chapter. 1, note 10.
7. See note 5.

8. C. G. Ehrenberg, *Die Infusionsthierchen als vollkommene Organismen.* Voss, Leipzig (1838).
9. R. Wagner, *Lehrbuch der vergleichenden Anatomie.* Voss, Leipzig (1834–5).
10. R. Wagner, *Icones Physiologicae.* Voss, Leipzig (1839).
11. W. Hewson, *Observations and Experiments of the Late William Hewson, F.R.S.* (M. Falconar, ed.). Royal Society, London (1777). This work was completed by G. Gulliver: *The Works of William Hewson, F.R.S.* Sydenham Society, London (1846). The first part of Hewson's work was presented to the Royal Society on 17 and 24 June 1773.
12. C. F. Heusinger, *System der Histologie.* Bärecke, Eisenach (1822).
13 See Chapter 3, note 10.
14. See Chapter 3, note 14.
15. J. Müller, *Handbuch der Physiologie des Menschen.* Hölscher, Coblenz (1833–8).
16. See Chapter 2, note 8.
17. J. Müller, *Vergleichende Anatomie der Myxinoiden, der Cyclostomen mit durchbohrten Gaumen.* Königliche Academie der Wissenschaften, Berlin (1835).

CHAPTER 6

1. C. Dobell, *Anthony van Leeuwenhoek and his Little Animals.* Bale & Danielsson, London (1932).
2. A. Schierbeck, *Measuring the Invisible World. The life and works of Antoni van Leeuwenhoek, FRS.* Abelard-Schuman, London and New York (1959).
3. See Chapter 1, note 16.
4. Ibid.
5. R. Hooke, *Lectiones Cutlerianae.* Royal Society, London (1679). Leeuwenhoek's letter concerning conjugation of animalcules is in Collection No. 3 and bears the date 10 December (1681).
6. A. van Leeuwenhoek, *Phil. Trans. Roy. Soc.* No. 213 (1694): 194.
7. Ibid., **23,** No. 283 (1703): 1304.
8. J. R. Baker, *Quart. J. microscop. sci.* **90** (1949): 87.
9. A. Trembley, *Phil. Trans. Roy. Soc.* **43** (1744): 169.
10. Ibid., **44** (1747): 627.
11. A. Trembley, Manuscript letter to Count Bentinck (1766) cited by J. R. Baker. *Quart. J. microscop. sci.* **94** (1953): 407.
12. L. Spallanzani, *Opusculi di fisica animale e vegetabile.* Società Tipografica Modena (1776).
13. L. Spallanzani, *Dissertazioni di physica animale e vegetabile.* Società Tipografica Modena (1780).
14. J. Senebier, *Observations et expériences faites sur les animalcules des infusions.* Geneva 1786. A reprint of this work was published by Gauthier-Villars, Paris in 1920.
15. O. F. Müller, *Animalcula infusoria fluviatilia et marina.* Möller, Hauniae (1786).
16. C. F.-A. Morren, *Annales des sciences naturelles* **20** (1830): 404.
17. Ibid., Seconde série **5** (1836): 257.
18. See Chapter 5, note 8.
19. C. G. Ehrenberg, *Organisation, Systematik und Geographisches Verhältnis der Infusionsthierchen.* Akademie der Wissenschaften, Berlin (1830).
20. C. G. Ehrenberg, *Zur Kenntnis der Organisation in der Richtung des kleinsten Raumes.* Akademie der Wissenschaften, Berlin (1832).
21. L. Oken, *Die Zeugung.* Goebhardt, Bamburg and Würzburg (1805).
22. L. Oken, *Lehrbuch der Naturphilosophie.* Fromann, Jena (1809); 3rd edn, Shulthess, Zürich (1843).
23. L. Oken, *Allgemeine Naturgeschichte für alle Stände.* Hoffmann, Stuttgart (1835).

CHAPTER 7

1. See Chapter 1, note 10.
2. H. Mohl, *Vermischte Schriften botanischen Inhalts.* Fues, Tübingen (1845).
3. H. Mohl, *Allgemeine botanische Zeitung* **1** (1837): 17.
4. H. Mohl, *Flora* **20** (1837): 1.
5. B. C. Dumortier, *Nova Acta Phys.-Med. Acad. Caesar. Leopold-Carolinae Nat. Curios* (part 1) **16** (1832): 217.

6. B. C. Dumortier, *Annales des sciences naturelles* (seconde série) **8** (1837): 129.
7. See Chapter 6, note 16.
8. See Chapter 5, note 5.
9. See note 3.
10. See note 2.
11. C. Nägeli, *Zeitschrift für wissenschaftliche Botanik* **1** (part 1), 36 (1844).
12. See Chapter 3, ref. 7.
13. B. C. Dumortier, *Annales des sciences naturelles* (seconde série) **8** (1837): 129.
14. See note 5.
15. See note 3.
16. See note 11.
17. See note 5.
18 See Chapter 4, note 7.
19. J. P. Vaucher, *Histoire des conferves d'eau douce*. Paschoud, Genève (1803).
20. See Chapter 4, note 4.
21. A. de Quatrefages, *Annales des sciences naturelles* (seconde série) **2** (Zoologie) (1834): 107.
22. See note 6.
23. J.-L. Prévost and J.-B. Dumas, *Annales des sciences naturelles* **2** (1824): 100, 129.
24. See note 3.
25. See Chapter 1, note 10.
26. See note 25.
27. A. Brogniart, *Annales des sciences naturelles* **12** (1827): 225.
28. See Chapter 4, note 18.
29. See note 2.
30. See Chapter 3, note 2.
31. H. Mohl, *Botanische Zeitung* **2** (1844): col. 273.
32. Ibid., **4**, (1846): col. 73.
33. F. Dujardin, *Annales des sciences naturelles* (seconde série) **4** (Zoology) (1835): 343.
34. G. Valentin, *Nova Acta Phys.-Med. Acad. Caesar. Leopold.-Carolinae. Nat. Curios* **18** (1836): 51.
35. See note 32.
36. See Chapter 10, note 19.
37. T. R. Jones, *A General Outline of the Animal Kingdom*. van Voorst, London (1841).
38. Dr Kützing, (initials not given), *Linnaea* **15** (1841): 546.
39. See note 9.

Chapter 8

1. See Chapter 6, note 5.
2. See Chapter 1, note 16.
3. See Chapter 5, note 4.
4. See Chapter 2, note 9.
5. F. Bauer, *Illustrations of Orchidaceous Plants*. Ridgeway, London (1830–38).
6. R. Brown, *Trans. Linn. soc. Lond.* **16** (1833): 685.
7. R. Wagner, *Arch. Path. Anat. wiss. Med.* (1835): 373.

Chapter 9

1. See Chapter 3, note 17.
2. See Chapter 2, note 8, and Chapter 3, note 4.
3. *Jan Evangelista Purkyně 1787–1869* (V. Kruta ed.). Centenary Symposium, Prague 1969. Acta Fac. Med. Univ. Brunensis. Brno (1971).
4. See Chapter 5, note 15.
5. J. E. Purkyně, *Symbolae ad ovi avium historiam ante incubationem* (Contributions to the developmental history of the bird's egg before incubation). 1st edn Vratislaviae (1825); 2nd edn Lipsiae (1830). Reprinted in *Opera selecta*, Tomus 1. Opera facultatis medicae universitatis carolinae pragensis (1948).
6. Quoted in *Purkyně* (See note 3): p. 77.
7. See note 2.
8. J. Henle, *Allgemeine Anatomie*. Voss, Leipzig (1841).
9. G. G. Valentin, *Repertorium Anat. Physiol* **1** (1836): 36.
10. G. G. Valentin, *Handbuch der Entwickelungsgeschichte*. Rücker, Berlin (1835).
11. G. G. Valentin, *Histogeniae plantarum atque animalium inter se comparatae specimen* (A comparative study of histogenesis in plants and animals), (E. Hintsche ed.). *Berner Beitr. Gesch. Med. und Naturw.* No. 20 (1963).
12. See Chapter 7, note 34.
13. G. G. Valentin, *Repertorium. Anat. Physiol* **4** (1839): 1.
14. See note 10.
15. R. Wagner, *Lehrbuch der Physiologie*. Leipzig (1842).
16. R. Heidenhain, 'Purkinje, Johannes

Evangelista', *Allgemeine deutsche Biographie* **26** (1888): 717.
17. J. E. Purkinje and G. G. Valentin, *Arch. Anat. Physiol. wiss. Med.* (1834): 391.
18. See chapter entitled 'Cililary Movement' in *Purkyně* (See note 3): p. 93.
19. J. E. Purkinje, *Arch. Anat. Physiol. wiss. Med.* (1836): 289.
20. J. E. Purkinje and G. G. Valentin, ibid. (1835): 159.
21. K. E. Rothschuh, 'Von der Histomorphologie zur Histophysiologie' in *Purkyně* (See note 3): p. 197.
22. J. E. Purkinje, *Bericht über die Versammlung deutscher Naturforscher und Aerzte in Prag in September 1837. Opera selecta* (1948): p. 109. (See note 5.)

Chapter 10

1. See Chapter 5, note 15.
2. See Chapter 5, note 17.
3. J. Müller, *Archiv. Anat. Physiol. wiss. Med.* (1838): 91.
4. J. Müller, *Ueber den feineren Bau und die Formen der krankhaften Geschwülste.* Reimer, Berlin (1838).
5. See Chapter 3, note 2.
6. M. J. Schleiden, *Grundzüge der wissenschaftlichen Botanik.* Engelmann, Leipzig (1842).
7. See Chapter 3, note 17.
8. T. Schwann, *Arch. Anat. Physiol. wiss. Med.* (1836): 90.
9. T. Schwann, *Poggendorfs Ann.* **41** (1837): 184.
10. L. Spallanzani, *Saggio di osservazione microscopiche concernenti il sistema della generazione dei Signori Needham e Buffon.* Modena (1765).
11. F. Schulze, *Poggendorfs Ann.* **39** (1836): 487.
12. T. Schwann, *Frorieps Neue Notizen* **93** (1838): 33; ibid., **103** (1838): 225; ibid., **107** (1838): 21.
13. See Chapter 3, note 17.
14. See Chapter 2, note 8, and Chapter 3, note 4.
15. G. Prochaska, *Bemerkungen über den Organismus des menschlichen Körpers.* Beck, Wien (1810).
16. J. Henle, *Symbolae ad anatomiam villorum intestinalium imprimis eorum epithelii et vasorum lacteorum.* Commentatio academica, Berlin (1837).
17. See Chapter 9, note 10.
18. G. G. Valentin, *Grundzüge der Entwickelung der tierischen Gewebe* in Wagner's *Lehrbuch.* See Chapter 9, note 15.
19. J. E. Purkinje, *Uebersicht der Arbeiten und Veränderungen der Schlesischen Gesellschaft für vaterländische Cultur im Jahre 1839* **16**. Breslau (1840): p.81.
20. J. E. Purkinje. *Jahrbücher für wissenschaftliche Kritik*, No. 5 (July) (1840): 33.

Chapter 11

1. K. B. Reichert, *Arch. Anat. Physiol. wiss. Med.* (1841): 162.
2. K. B. Reichert, *Das Entwickelungsleben im Wirbelthier-Reich.* Hirschwald, Berlin (1840).
3. K. B. Reichert, *Arch. Anat. Physiol. wiss. Med.* (1841): 523.
4. Studnička (See Chapter 2, note 8): 151.
5. J. Henle, Article on *Galle* in *Encyclop. Wörterbuch d. med. Wiss* **13** (1835).
6. See Chapter 10, note 6.
7. J. Henle, *Arch. Anat. Physiol. wiss. Med.* (1838): 103.
8. See Chapter 9, note 8.
9. A. Reyer, *Leben und Wirken des Naturhistorikers Dr Franz Unger.* Leuchner & Lubensky, Graz (1871).
10. F. Unger, *Flora* **45** (1832): 713.
11. F. Unger, *Nova Acta Phys.-Med. Acad. Caesar. Leopold–Carolinae Nat. Curios* **14** (1850): 509.
12. A. Ecker, *Zeitschrift für wissenschaftliche Zoologie* **1** (1848): 218.
13. F. Unger, *Die Anatomie und Physiologie der Pflanzen.* Hartleben, Pest, Wien and Leipzig (1855).
14. F. Unger, *Linnaea* (1841): 385.
15. F. Unger, *Botanische Zeitung* **2** (1844): Cols 489, 506, 521.
16. See Chapter 7, note 11.
17. C. Nägeli, *Zsch. f. wissen. Bot.* **3** (1846): 22.
18. L. C. Dunn, Introduction to *Genetics in the 20th Century* (ed. L. C. Dunn). Macmillan, New York (1951).

19. V. Orel, *Gregor Mendel, the First Geneticist*. Oxford University Press, Oxford (1996).
20. R. Remak, *Med. Z. Ver. Heilk. Pr.* **10** (1841): 127; *Canstatts Jahresber. ges. Med.* **1** (1841): 17.
21. R. Remak, *Frorieps Neue Notizen* **35** (1845): 305.
22. R. Remak, *Arch. Anat. Physiol. wiss. Med.* (1852): 47.

CHAPTER 12

1. R. de Graaf, *De mulierum organis generationi inservientibus tractatus novus.* Ex officina Hackiana Lugduni Batav, Lyons (1672). Facsimile by J. A. van Dongen (1965).
2. L. W. Bischoff, *Entwicklungsgeschichte des Kaninchen-Eies.* Vieweg, Braunschweig (1840).
3. K. E. von Baer, *De ovi mammalium et hominis genesi.* Leipzig (1827).
4. M. D. Roffredi, *Observations et mémoires* **5** (1775): 197.
5. See Chapter 6, note 13.
6. See Chapter 7, note 23.
7. D. M. Rusconi, *Développement de la grenouille commune depuis le moment de sa naissance jusque à son état parfait.* Giusti, Milan (1826).
8. K. E. von Baer, *Arch. Anat. Physiol. wiss. Med.* (1834): 481.
9. M. Rusconi, *Arch. Anat. Physiol. wiss. Med.* (1836): 205.
10. Ibid., (1836): 278.
11. A. de Quatrefages, *Annales des sciences naturelles* Seconde série (Zoologie), **2** (1834): 107.
12. C. T. von Siebold, 'Zur Entwickelungsgeschichte der Helminthen' in K. F. Burdach, *Die Physiologie als Erfahrungswissenschaft*, Vol. 3. Voss, Leipzig (1835): 32.
13. J. V. Coste, *Frorieps Neue Notizen* **38** (1835): Col. 241.
14. M. Barry, *Phil. Trans. Roy. Soc.* **128** (1838): 301.
15. Ibid., **129** (1839): 307.
16. Ibid., **130** (1840): 529.
17. Ibid., **131** (1841): 193.
18. Ibid., **131** (1841): 195.
19. Ibid., **131** (1841): 217.
20. C. Bergmann, *Arch. Anat. Physiol. wiss. Med.* (1841): 89.
21. Ibid., (1842): 92.
22. C. Vogt, *Untersuchungen über die Entwicklungsgeschichte der Geburtshelferkroete.* Jent & Gassman, Solothurn (1842).
23. H. Bagge, *Dissertatio inauguralis de evolutione Strongyli auricularis et Ascaris acuminatae viviparorum.* Ex officina Barfusiana, Erlangae (1841).
24. H. Rathke, *Frorieps Neue Notizen* **24** (1842): 160.
25. See Chapter 11, note 2.
26. See Chapter 11, note 3.
27. K. B. Reichert, *Arch. Anat. Physiol. wiss. Med.* (1846): 196.
28. R. Remak, *Untersuchungen über die Entwickelung der Wirbelthiere.* Reimer, Berlin (1855).

CHAPTER 13

1. H.-P. Schmiedebach, *Robert Remak (1815–1865).* Gustav Fischer, Stuttgart. (1995).
2. A. Wrzosek, ibid., pp. 4, 354.
3. V. H. H. Green, *The Commonwealth of Lincoln College, 1427–1977* Oxford University Press, Oxford (1979).
4. See Chapter 12, note 28.
5. See Chapter 11, note 22.
6. See Chapter 11, note 20.
7. See Chapter 11, note 21.
8. R. Remak, *Deutsche Klinik* **6** (1854): 170.
9. See Chapter 12, note 28.
10. B. Kisch (Article on Robert Remak) 'Forgotten Leaders in Modern Medicine', *Trans. Amer. Philos. Soc.* (New Series) **44** (1954): 227.
11. R. Virchow, *Arch. path. Anat. Physiol. klin. Med.* **1** (1847): 207.
12. Ibid., (1851): 197.
13. R. Virchow, *Canstatts Jahresber. ges. Med.* **2** (1854): 11.
14. R. Virchow, *Arch. path. Anat. Physiol. klin. Med.* **8** (1855): 3.
15. R. Virchow, *Cellularpathologie.* Hirschwald, Berlin (1871).
16. R. Remak, *Arch. Anat. Physiol. wiss. Med.* (1858): 178.
17. Ibid., (1862): 230.

Chapter 14

1. K. B. Reichert, *Arch. Anat. Physiol. wiss. Med.* p. 1 (1847).
2. See Chapter 11, note 22.
3. See Chapter 12, note 28.
4. See Chapter 13, note 16.
5. See Chapter 7, note 11.
6. R. Virchow, *Arch. path. Anat. Physiol. wiss. Med.* **11** (1857): 89.
7. R. Breuer, *Melemata circa evolutionem ac formas cicatricum*. editio altera. Schumann, Vratislaviae (1844).
8. F. Günzburg, *Studien zur speciellen Pathologie*. Brockhaus, Leipzig (1848).
9. See Chapter 12, note 24.
10. C. Gegenbaur, *Abhand. Naturforsch. Gesell. zu Halle* **4** (1858): 1.
11. A. Kölliker, *Entwickelungsgeschichte der Cephalopoden*. Meyer and Zeller, Zürich (1844).
12. K. E. von Baer, *Bull. Acad. Imp. St Pétersbourg* (Classe phys.-math.) **5** (1846): col. 231.
13. M. Schultze, *Arch. Anat. Physiol. wiss. Med.* (1861): 1.
14. A. Weismann, *Zeit. wiss. Zool.* **13** (1863): 159.
15. A. Krohn, *Arch. Anat. Physiol. wiss. Med.* (1852): 312.
16. J. von Hannstein, *Sitz. niederrhein. Ges.* Bonn (1870): 217.
17. See Chapter 9, note 8.
18. C. Nägeli, *Zur Entwickelungsgeschichte des Pollens bei den Phanerogamen*. Ovell & Füssli, Zürich (1842).
19. A. Kowalevski, *Mém. Acad. Imp. St. Pétersbourg*, 7me série **16** (1871): 1.
20. See Chapter 13, note 16.
21. W. Hofmeister, *Bot. Zeit.* **6** (1848): cols. 425, 649, 670.
22. W. Hofmeister, *Die Entstehung des Embryos der Phanerogamen*. Friedrich Hofmeister, Leipzig (1849).
23. See note 21.
24. See note 21.
25. W. Hofmeister, *Die Lehre von der Pflanzenzelle*. Engelmann, Leipzig (1867).
26. W. Flemming, *Arch. mikroscop. Anat.* **16** (1879): 302.
27. See note 19.
28. E. Russow, *Mém. Acad. Imp. St. Pétersbourg*, 7me série. **19** (1872): 1.
29. A. Schneider, *Ber. Oberhess. Ges. Natur-und Heilk.* **14** (1873): 69.
30. J. Tschistiakoff, *Bot. Zeit.* (1875): **33** cols 1, 17, 80.
31. T. von Ewetsky, *Untersuch. a.d. path. Institut in Zürich* **3**, (1875): 89.
32. E. Strasburger, *Zellbildung und Zelltheilung*. Hermann Dabis, Jena (1875).
33. O. Bütschli, *Zeit. wiss. Zool.* **25** (1875): 201.
34. O. Bütschli, *Abhand. Senckenberg. Naturforsch. Ges.* **10** (1876): 213.
35. W. Mayzel, *Centralbl. med. Wiss.* **50** (1875): 849.
36. C. J. Eberth, *Arch. path. Anat. Physiol. klin. Med.* **67** (1876): 523.
37. W. Mayzel, *Centralbl. med. Wiss. Year* **15** (1877): 196.
38. E. G. Balbiani, *C. R. Acad. Sci. Paris* **83** (1876): 831.

Chapter 15

1. See Chapter 11, note 12.
2. See Chapter 11, note 1.
3. A. de Bary, *Zeit. wiss. Zool.* **10** (1860): 88.
4. A. de Bary, *Die Mycetozoen (Schleimpilze)* Engelmann, Leipzig (1864).
5. M. Schultze, *Arch. Anat. Physiol. wiss. Med.* (1858): 330.
6. M. Schultze, *Arch. Naturgesch.* **26** (1860): 287.
7. M. Schultze, *Arch. Anat. Physiol. wiss. Med.* (1861): 1.
8. E. Haeckel, *Allgemeine Anatomie der Organismen*. Reimer, Berlin (1866).
9. E. Haeckel, *Die Radiolarien (Rhizopoda radiaria)*. Reimer, Berlin (1862).
10. J. von Hannstein, Article on protoplasm in *Sammlung von Vorträgen für das deutsche Volk* (W. Frommel and F. Pfaff, eds). Winter, Heidelberg (1880): p. 125.
11. H. de Vries, *Jahrb wiss. Bot.* **14** (1884): 427.
12. N. Pringsheim, *Untersuchungen über den Bau und die Bindung der Pflanzenzelle (1854)* repr. in *Gesammelte Abhandlungen von N. Pringsheim* **3** (1896): 33.

13. C. Nägeli, *Primordialschlauch und Diosmose (Endosmose und Exosmose) der Pflanzenzelle* in *Pflanzenphysiologische Untersuchungen von C. Nägeli und C. Cramer* (1855).
14. E. Overton, *Vierteljahrschr. naturf. Gesell.* Zürich **40** (1895): 159. **44** (1899): 88. *Jahrb. wiss. Bot.* **34** (1900): 669.

Chapter 16

1. E. G. Balbiani, *J. physiol.* **4** (1861): 102, 481, 465.
2. See Chapter 14, note 34.
3. See Chapter 14, note 38.
4. E. van Beneden, *Bull. Acad. roy. Belg.* **40** (1875): 686.
5. W. S. Sutton, *Biol. Bull.* **4** (1902): 124.
6. See Chapter 14, note 32.
7. E. Strasburger, *Jena. Zeitsch. f. Naturwiss.* **11** (1877): 435.
8. Ibid., **13** (1879): 93.
9. See Chapter 14, note 32 (3rd edition: 1880).
10. W. Flemming, *Schrift. naturwiss. Ver. Schleswig-Holstein* **3** (1878): 23.
11. W. Flemming, *Arch. path. Anat. Physiol. klin. Med.* **77** (1879): 1.
12. See Chapter 14, note 26.
13. Professor Peremeschko, *Zentralbl. med. Wiss.* **16** (1878): 547.
14. W. Schleicher, ibid., **16** (1878): 418.
15. W. Schleicher, *Arch. mikroskop. Anat.* **16** (1879): 248.
16. W. Flemming, ibid., **13** (1877): 693.
17. See note 10.
18. See note 11.
19. See Chapter 14, note 26.
20. W. Flemming, *Arch. mikroskop. Anat.* **18** (1880): 151.
21. Ibid., **20** (1881): 1.
22. W. Flemming, *Zellsubstanz Kern und Zelltheilung.* Vogel, Leipzig (1882).
23. N. Warneck, *Bull. Soc. Imp. Nat. Moscou* **23** (1850): 90.
24. O. Bütschli, *Nova Acta Phys. Med. Acad. Caesar. Leopold. Carolinae Nat. Curios* **36** (1873): 1.
25. R. B. Goldschmidt, 'Otto Bütschli (1848–1920) Pioneer of Cytology', in *Science Medicine and History. Essays in honour of Charles Singer* Vol. 2. Oxford University Press, Oxford (1953), p. 223.
26. See note 24.
27. H. Fol, *C. R. Acad. Sci. Paris* **83** (1876): 667; **84** (1877): 268; **84** (1877): 357.
28. See Chapter 14, note 34.
29. H. Fol, *C. R. Acad. Sci. Paris* **84** (1877): 357.
30. O. Hertwig, *Morphol. Jahrb.* **1** (1875): 347.
31. L. Auerbach, *Zur Charakteristik und Lebensgeschichte der Zellkerne. Organologische Studien* Vol. 1. Morgenstern, Breslau (1874).
32. O. Hertwig, *Morphol. Jahrb.* **3** (1877): 271; **4** (1878): 156.
33. E. Selenka, *Zoologische Studien. Befruchtung des Eies von Toxopneustes variegatus.* Engelmann, Leipzig (1878).
34. E. van Beneden, *Arch. Biol.* **4** (1883): 265.
35. Ibid.
36. E. van Beneden and A. Neyt, *Bull. Acad. roy. Belg.*, (Sér. 3.) **14** (1887): 215.
37. See note 34.
38. A. Schneider, *Das Ei und seine Befruchtung.* Breslau (1883).
39. E. Heuser, *Bot. Zentralbl.* **17** (1884): 27, 57, 85, 117, 154.
40. C. Rabl, *Morphol. Jahrb.* **10** (1885): 24.
41. See note 33.
42. E. Mayr, *The Growth of Biological Thought.* Harvard University Press, Cambridge, Mass. (1982).
43. See note 15.
44. See note 22.
45. E. Strasburger, *Arch. mikroskop. Anat.* **23** (1884): 246.
46. W. Waldeyer, ibid. **32** (1888) 1.
47. M. Heidenhain, ibid., **43** (1894): 423.
48. H. Lundegårdh, *Archiv. Zellforsch.* **9** (1913): 205.

Chapter 17

1. E. Haeckel, *Generelle Morphologie der Organismen: Allgemeine Grundzüge der organischen Formen-Wissenschaft.* Reimer, Berlin (1866).
2. J. G. Kölreuter, *Vorläufige Nachricht von einigen das Geschlecht der Pflanzen betreffenden Versuchen und Beobachtungen* (1761) *Fortsetzung*

(1763) *Zweyte Fortsetzung* (1764) *Dritte Fortsetzung* (1766). Gleditschische Buchhandlung, Leipzig.
3. C. Nägeli, *Mechanisch-physiologische Theorie der Abstammungslehre*. Oldenbourg, Munich and Leipzig (1884).
4. W. Roux, 'Ueber die Bedeutung der Kerntheilungsfiguren', in *Gesammelte Abhandlungen über Entwickelungsmechanik der Organismen*. Engelmann, Leipzig (1883): p. 125.
5. E. Strasburger, *Neue Untersuchungen über den Befruchtungsvorgang bei den Phanerogamen als Grundlage für eine Theorie der Zeugung*. Fischer, Jena (1884).
6. A. D. Hershey and M. Chase, *J. Gen. Physiol.* **36** (1952): 39.
7. O. Hertwig, *Z. Naturwiss.* **18** (1884): 21.
8. Ibid. **18** (1885): 276.
9. F. Miescher, *Histochemische und physiologische Arbeiten*. (W. His. ed.), Vogel, Leipzig (1897).
10. O. T. Avery, C. M. Macleod and M. McCarty, *J. exp. Med.* **79** (1944): 137.
11. A. Weismann, *Über die Vererbung*. Fischer, Jena (1883); *Die Kontinuität des Keimplasmas als Grundlage einer Theorie der Vererbung*. Fischer, Jena (1885).
12. T. Boveri, *Sitzungsber. Ges. Morph. Physiol. München* **3**, (1887): 153; *Z. Naturwiss.* **22**, (1888): 685.
13. Ibid., **21** (1887): 423.
14. See Chapter 16, note 38
15. T. Boveri, *Verh. phys.-med. Ges. Würzburg*. **35** (1902): 67.
16. T. Boveri, *Verh. zool. Ges.* (Versamml. Würzburg) **13** (1903): 10.
17. T. Boveri, *Ergebnisse über die Konstitution des chromatischen Substanz des Zellkerns*. Fischer, Jena (1904); *Z. Naturwiss* **43** (1907): 1.
18. R. Fick, *Verh. anat. Ges. Tübingen. Anat. Anz.* **94** (1899): 16; *Arch. Anat. Phys. (Anat. Abt.)* Supplement (1905); *Ergeb. Anat. Entwicklungsgesch.* **16** (2 Abteil; 1907): 1.
19. T. Boveri, *Arch. Zellforsch.* **3** (1909): 181.
20. See note 15.
21. See note 16.
22. T. Boveri, *Ergebnisse über die Konstitution*.
23. T. H. Montgomery,. *Trans. Amer. Philos. Soc.* **20** (1901): 154.
24. C. E. McClung, *Biol. Bull.* **3** (1902): 43.
25. W. S. Sutton, *Biol. Bull.* **4** (1902): 124.
26. Ibid., **4** (1903): 231.
27. T. Boveri, *Zur Frage der Entstehung maligner Tumoren*. Fischer, Jena (1914).
28. T. H. Morgan, *Am. Nat.* **44** (1910): 449; *Science* **32** (1910): 120.
29. T. H. Morgan, C. B. Bridges and A. H. Sturtevant, *The Genetics of Drosophila*. Nijhoff, 's Gravenhage (1925).
30. T. S. Painter, *Science* **78** (1933): 585.
31. T. S. Painter, *J. Hered.* **25** (1934).

Appendix

Chapter 1

p. 9 Microscopium ostendit, esse parenchyma libri nihil aliud, praeter massam infinitam cellularum parvarum vel bullarum fixatarum.

Chapter 2

p. 16 Je voudrais bien scavoir quelle foy on adjoute chez vous aux observations de nostre Monsieur Leeuwenhoek qui convertit toute chose en petites boules.

p. 18 Fibra enim physiologio id est, quod linea geometrico, ex qua nempe figurae omnes oriuntur.

p. 20 Crescunt igitur folia vesiculis novis, veteribus interpositis, maxima ex parte, partim vero quoque distensione vesicularum. Et rami et trunci amplificantur aeque, interpositione novorum vasorum interea, quae hactenus illos constituerunt, partim, et partim veterum vasorum dilatatione.

p. 20 Folium recentius, ex gemma sumtum, se monstrat compositum ex meris vesiculis, fibris vero, vasculis, striis quibuscumque plane destitutum.

p. 20 . . . ex pura adhuc, aequabili vitrea substantia constant, absque ullo vesicularum, vel vasorum vestigio.

p. 21 Partes constitutinae, ex quibus omnes corporis animalis partes in primis initiis componuntur, sunt globuli, mediocri microscopio cedentes semper. Quis autem diceret, se non potuisse corpus videre propter exiguitatem, cuius tamen particulae constituentes propter exiguitatem ipsum fugere nescirent?

pp. 21–2 Les cylindres tortueux primitifs que j'ai trouvés dans le tissu celluleux des nerfs, des tendons et des muscles, sont de toutes les parties ou organes que je connaisse, les plus petites. Ils le sont beaucoup plus que les plus petits vaisseaux rouges qui ne laissent passer qu'un globule de sang à la fois. Toutes les tentatives que j'ai faites pour les décomposer en cylindres moindres . . . ont été inutiles.

p. 22 . . . je ne connais dans le corps animal aucune partie qui, ayant du tissu celluleux, ne présente pas les cylindres tortueux.

Chapter 3

p. 23 Auf Raspails Arbeit mich einzulassen, scheint mir mit der Würde der Wissenschaft unverträglich. Wer Lust dazu verspürt, mag sich an ihn selbst wenden.

p. 23 On croit généralement que la théorie cellulaire est une conception allemande; c'est une complète erreur. Elle n'est née ni en 1838 ni même en 1837; elle n'est fille ni de Schwann ni de Schleiden. Elle est plus vieille de douze ans; elle est française et appartient à M. Raspail.

p. 24 Par l'effet des propositions assemblées autour du concept de cellule, les théories de Dutrochet et de Raspail avant celle de Schwann, pouvaient exercer une fonction programmatique dès la décennie 1820.

p. 26 Il me suffira de rappeler que partout sa structure intime m'a paru identique, et ses globules élémentaires semblables, par leur forme et par leur diamètre, à ceux que l'on voit nager dans le pus, dans le lait, etc.

p. 26 . . . un nombre plus ou moins grand de ces corpuscles, dont la nature chimique peut différer, mais dont la forme et probablement le volume ne varient que peu.

p. 27 Ces globules, que l'on peut appeler élémentaires, sont peut-être formés à leur tour d'autres corpuscules plus petits et que nos moyens d'investigation ne nous ont point permis d'apercevoir.

p. 28 *La vie est une,* les différences que présentent ses divers phénomènes, chez tous les êtres qu'elle anime, ne sont point des différences fondamentales; lorsqu'on poursuit ces phénomènes jusqu'à leur origine, on voit les différences disparaître et une admirable uniformité de plan se dévoile.

p. 29 Cette uniformité de structure intime prouve que les organes ne diffèrent véritablement entre eux que par la nature des substances que contiennent les cellules vésiculaires dont ils sont entièrement composés: c'est dans les cellules que s'opère la sécrétion du fluide propre a chaque organe. . . .

p. 29 Cet organe étonnant [the cell] par la comparison que l'on peut faire de son extrême simplicité avec l'extrême diversité de sa nature intime est véritablement la pièce fondamentale de l'organisation; tout, en effet, dérive évidemment de la cellule. . . .

p. 32 Donnez-moi une vésicule organique douée de vitalité, et je vous rendrai le monde organisé.

p. 35 . . . parce qu'elle est contenue dans des vésicules mères, qui ont commencé par être elle-mêmes de la Globuline, offre les mêmes caractères que les deux premières: mêmes formes, mêmes couleurs, même mode de reproduction. Mais elle se distingue en ce qu'au lieu de vivre et de croître separément, elle reste dans l'intérieur des vésicules-mères, où gênée dans son développement, elle perd sa forme globulineuse, devient plus ou moins hexagonale, se soude ou s'entregreffe par ses surfaces, et constitue une nouvelle masse de tissu cellulaire.

pp. 35–6 Tout corps propagateur végétal doit son origine à une vésicule favorisée qui lui sert de mère et de conceptacle soit que cette vésicule appartienne à une Globuline solitaire ou à une Bichatie, soit à un végétal confervoide, soit à un grain de pollen, soit enfin à celles *associées* d'une masse de tissu cellulaire d'un végétal d'ordre plus élevé.

p. 36 Un grain de Globuline, produit par extension des parois intérieures de l'une des vésicules conceptables du tissu cellulaire, est l'origine ou le germe propagateur, soit des vésicules futures d'un nouveau tissu cellulaire, soit de tout corps capable de propager l'espèce.

Chapter 4

p. 37 La première idée, l'idée fondamentale est que toute l'organisation végétale est formée par un *seul et même tissu membraneux*, différemment modifié. Ce fait est la base de tous les autres.

p. 37 Je pars de ce principe que la masse entière de la plante est un tissu cellulaire dont les loges diffèrent par leurs formes et leur dimensions. Cette idée simple est la base de toute ma théorie.

p. 37 Je ne vois qu'une exception à cette loi; elle a lieu pour les *trachées*. Ces lames étroites, roulées en hélice comme un tire-bourre.

p. 38 . . . et je reconnus enfin, que cette union, qui ne me semblait expliquable que par l'existence de fibres latérales, provient de ce que la masse entière du végétal n'est qu'un tissu membraneux lequel offre des vacuosités variables dans leurs formes et leurs dimensions; de telle sorte, que les unes sont de petites cellules régulières ou irrégulières, et les autres, des tubes peu ou moins prolongés.

pp. 37–8 Der erste Anfang aller Organisation des Lebendigen ist ein Aggregat von Bläschen, die unter einander keine Verbindung haben. Aus diesen entstehen alle lebende Körper, so wie auch alle darin wieder aufgelöst werden.

p. 39 Das Erste was an dem Saamenkorne und Ey sich bildet, ist eine doppelte äussere Hülle, von welchen die aüssere härtere den Namen des Chorion, die innere zartere den des Amnion enhalten hat. Diese Membranen zeigen sich schon, wenn das Innere des Saamenkorns und Eys noch eine flüssige Substanz ohne sichtbare Organisation ist.

p. 40 Meine Meinung von der Entstehung der Blasen, welche in ihrer Gesammtheit das Zellgewebe ausmachen, aus den Körnern welche man in den Zellen findet, ist nach Mirbels Aussprüche ein Gespinst der Einbildungskraft. Gerechter ist Link, in dem er sie bezweifelt und die Gründe seines Zweifels angibt. So wenig entscheidend diese sind, so wenig bin ich geneigt, jener Meinung die überredende Kraft der Wahrheit beyzumessen: es ist und bleibt vielmehr nur eine wahrscheinliche Vermuthung.

p. 40 . . . wir haben hier also einen nicht wohl zu bestreitenden Beweis vor Augen, dass das körnige Wesen die Materie hergebe woraus die Zellen und Fasern des jungen Pflänzchens gebildet werden.

p. 41 Es macht keinen grossen Unterschied, ob es bey diesem Uebergänge seine körnige Gestalt behält oder in eine gleichförmige Flüssigkeit aufgelöset wird, welches letzere das Wahrscheinlichere ist.

p. 42 Brisseau Mirbel, ein neuer Beobachter, behauptet deutliche Oeffnungen, als runde Löcher in den Wänden der Zellen gesehen zu haben. Kein anderer Beobachter bestätigt dieses.

p. 43 Man darf sich darüber nicht wundern; auch am thierischen Körper dringen sehr häufig Feuchtigkeiten durch jene unsichtbare Oeffnungen.

p. 43 D'abord, on ne peut se dissimuler que la *membrane végétale* ne soit percée d'une innombrable quantité de *pores* imperceptibles, qui favorisent le mouvement des fluides.

p. 43 Offenbar entsteht neues Zellgewebe zwischen den ältern Zellen. In den Zwischenräumen, wo man später die einfachen Zellengänge sieht, bemerkt man im jugendlichen Zustande dunkel gefärbte, wie aus einer zusammengedrängten Masse gebildete Streifen, die nicht selten ein äusserst feines Gewirre von Fasern und anderen kaum zu erkennenden Theilen entdecken lassen.

p. 43 Die erste Bildung des Organischen geschieht aus dem in organischen Behältern aufbewahren Flüssigen durch eine Kraft, welche Blumenbach passend Bildungstrieb genannt hat, deren Gesetze wir aber nicht kennen.

pp. 43–4 Der gebildete Theil wächst, in dem überall die feinsten unsichtbaren Theilchen eingeschoben werden. Eben so entwickeln sich neue Theile in den Zwischenräumen der ältern, und das Alte unterscheidet sich von dem Jungen nur in Menge und Grösse der einzelnen Theile.

p. 44 Unter Zellgewebe verstehen wir eine Sammlung mit einander verbundenen, von zarten Häuten geschlossenen Behälter, von verschiedenen, meist eckigen Gestalt, in welchen oft Säfte, oft aber blossen Luft enthalten ist.

p. 44 "Die Wände der Zellen sind die zartesten Häute, mehrenteils ohne sichtbare Oeffnungen oder Poren.

Chapter 5

p. 47 Wir wissen daher nicht ob, wie Gallini Platner und Ackermann angenommen haben, alle Theile des Körpers, und also auch die kleinsten Fasern und Blättchen aus einer schwammigen, d.h. von Zwischenräumen unterbrochenen Substanz bestehen, so dass also das schwammige oder zellige Gefüge die Grundform der thierischen Substanz wäre, oder ob die kleinsten Körnchen, Fasern und Blättchen vielmehr aus einer gleichartigen, nicht weiter in kleinere Theilchen getheilten, noch durch Form und Zwischenräume unterbrochenen Materie bestehen.

p. 48 Eben so wenig besitzt man hinreichende Beobachtungen darüber, wie bei der Bildung des menschlichen Embryo jene verschieden gestalteten kleinen Teilchen nach und nach entstehen, und manche Anatomen vermuthen nur, dass sich alle jene kleinen Theile aus ungeformter Materie und aus Kügelchen bildeten, die beide sogleich anfangs in der Materie vorhanden waren, aus der der Embryo entstehe; dass nämlich manche Kügelchen hohl würden, und Zellen bildeten, dass hierauf aus an einander gereiheten und vereinigten hohlen Kügelchen Röhren entständen u.s.w. In der That kann man sich mehrere Fälle als möglich denken.

p. 49 Im Jahre 1832 beobachtete Herr Dumortier die Vermehrung der Zellen durch wirkliche Theilung; sobald nämlich die Endzellen der *Conferva aurea* bedeutend länger geworden sind, als die darauf folgenden Zellen, so bildet sich in ihrem Innern eine Scheidewand, wodurch aus der einen Zelle jedesmal zwei hervorgehen. Später wurde eine solche Vermehrung der Zellen durch Theilung von H. Morren an *Closterien* und von H. Mohl an *Conferva glomerata* beobachtet; und gegenwärtig ist die Zahl solcher Beobachtungen sehr vergrössert worden.

p. 50 Am auffallendsten erscheint die Sporenbildung bei der Gattung *Marchantia*, wo man sich vollständig überzeugen kann dass dieselbe nicht innerhalb sogenannter Mutterzellen vor sich geht, sondern, wie bei anderen Laub-und Lebermoosen durch Selbsttheilung.

p. 50 Dergleichen Ansichten müssen aber den direkten Beobachtungen der neuesten Zeit weichen, und ich glaube im Folgenden beweisen zu können, dass die Vermehrung durch Selbsttheilung eine Erscheinung ist, welche gerade den Pflanzen in einem ausgedehnterem Grade zukommt, als den Thieren.

p. 50 Der Anbau dieses interessanten Feldes (Entwickelung der Gesetze in der organischen Formbildung) liegt seit Leeuwenhoek, Malpighi und Haller brach und ist erst in der allerneuesten Zeit wieder begonnen worden. Die grosse Verbesserung der Mikroskope und das wachsende Interesse jüngerer Beobacter wird in kurzer Zeit hier weiter aufhellen.

p. 52 Sie entstehen als weiche Tröpchen von plastischer Lymphe (Bildungsgewebe), diese nehmen, wenn es die Umgebungen gestatten, eine ganz kugelförmige Gestalt an; wenn dies die Umgebungen nicht gestatten, so werden es Scheiben, oder mehr eckige Körper; sie werden bald härter und bekommen ganz das Ansehen von Knorpeln, weswegen ich sie auch Chondroiden genannt habe.

p. 52 Die Zellen sind unregelmässig, einander ungleich, gleichen aber einigermassen den Pflanzenzellen darin, dass die Wände allseits geschlossen zu sein scheinen und meist in geraden Linien an einander stossen, so dass unregelmässig vieleckige Figuren auf den Durchschnitten sich zeigen.

Chapter 6

p. 59 Quand j'aperçus pour la première fois mon petit végétal, il se présenta sous la forme d'une croix de Malte fort réguliere.

p. 59 . . . l'individualité ne réside pas dans un assemblage d'êtres, mais dans chacun de ces êtres en particulier.

p. 59 Cellules primaires qui se divisent en quatre, et qui parviennent, par nuances insensibles, aux differens états de jeunes *Crucigènes.*

p. 60 de toute la périphérie du plus petit diamètre médian de la Clostérie s'était étendue vers le centre une lame circulaire qui séparait comme une cloison transparente les masses de chromules polarisées . . . La cloison se forme donc ici commes dans les *Conferves*, selon la belle observation de M. Dumortier. . . .

p. 60 Après que la membrane médiane est ainsi formée, l'articulation extérieure se manifeste sous la figure d'un trait noir circulaire qui limite exactement la base commune des deux cônes dont la Clostérie se compose. Ce trait est l'indice de la *déhiscence* qui s'opera plus tard sur la plante. . . .

p. 61 Die Gattung der Theilmonaden zeichnet sich durch unvollkommene Abschnürung der Individuen bei der Selbsttheilung von Traubenmonaden aus.

p. 61 Freiwillige Queertheilung und Längstheilung aber sind sehr in die Augen fallend.

p. 61 Der Fortpflanzung-Apparat der Monaden . . . besteht aus sehr vielen im ganzen Körper verstreuten, netzartig verbundenen Körnchen und aus einem verhältnissmässig grossen kuglichen und drüsigen Körper welcher sich bei der Selbstheilung mit theilt. Diese drüsige Kugel ist . . . offenbar einer männlichen Samendrüse ganz analog, und jene Körnchen sind Eier ganz ähnlich.

p. 61 Was den männlichen Theil des Sexualsystems anlangt, so ist die der Formen keineswegs ein Hinderniss für dessen Darstellung geblieben, ja er ist schon bei mehreren Arten deutlichgeworden.

p. 61 Der weibliche Theil zeigt sich als farbige, gleichgrosse, sehr zahlreiche Körnchen, die Eier; der männliche Theil bildet 1–2 rundliche Drüsen, die sich meist sehr auszeichen und einzelne contractile Blasen.

Chapter 7

p. 66 Le développement des conferves est aussi simple que leur structure, il s'opère par l'addition de nouvelles cellules aux anciennes et cette addition se fait toujours par l'extremité. La cellule terminale s'allonge plus que les inférieures (Voyez planche 10, fig. 15a), alors il s'opère dans le fluide intérieur une production médiane de la paroi interne qui tend à diviser la cellule en deux parties dont l'inférieure reste stationaire (Voyez planche 10, fig. 15b) tandis que la terminale s'allonge de nouveau, produit encore une nouvelle cloison intérieure et ainsi de même. La production de la cloison médiane est-elle dans le principe double ou simple, voilà ce qu'il est impossible de déterminer, mais toujours est il vrai de dire que plus tard elle parait double dans les conjugées (Voyez planche 11, fig. 34e) et lorsque deux cellules se séparent naturellement, elles sont chacune close aux extremités. C'est ce qui est facile a démontrer, pour les conferves, en les observant à leur maturité, et pour le tissu cellulaire lorsqu'il a subi l'influence de la gelée; dans cet état, les cellules continuent à renfermer les fluides qu'elles contenaient précédemment, ce qui ne serait pas si elles n'étaient closes par une membrane.

Ce fait de la production d'une cloison médiane dans les conferves nous parait expliquer bien clairement l'origine et le développement des cellules qui est jusqu'ici restée sans explication. . . .

p. 68 L'étude des infiniment simples de la création est une anatomie toute faite, et autant plus certain qu'elle montre à découvert ce que les êtres composés nous cachent dans leur intérieur. Nous avons vu que la formation des cellules des conferves s'opère par la production d'une cloison médiane; mais cette formation s'opère sur une seule ligne. Il ne se fait pas parmi les cellules aucune agglomération laterale, aucun point de réunion, aucun centre organique, mais elles se déposent en série linéare, et en se développant seulement et toujours par l'extremité, elles suivent la loi de l'élongation infinie. Ici encore les conferves nous montrent à découvert ce que les végétaux supérieurs nous cachent dans leur intérieur. Toute production de vaisseaux ou de fibres, toute série de cellules, suit la même loi qui paraît des mousses et des Jongermannes, présentent le même caractère; seulement les cellules au lieu d'être unisériées comme dans les conferves, présentent souvent des réunions de séries plus ou moins considérables.

p. 68 L'opinion de M. Kieser est donc inadmissible; d'ailleurs cette opinion, ainsi que celle de M. Tréviranus, est basée sur un hypothèse qui n'est rien moins que démontrée, celle de la transformation des grains ou corpuscles globuleux en cellules; et nous croyons pouvoir assurer d'après nos propres observations, que cette transformation n'a jamais lieu, et que les grains amylacés, ainsi que les corpuscules globuleux, sont des organes entièrement differens des cellules. Au contraire la production médiane d'une paroi interne a quelque chose de si analogue du reste de l'organisation, qu'on ne peut s'empêcher de l'admettre.

p. 69 "Presqu'au même instant et dans le même jour, ou au moins dans la même semaine, tous les grains de la *conferva jugalis* (j'en avais plusieurs milliers) s'ouvrirent par une de ces extremités comme les deux cotylédons d'une graine dont l'embryon se développe; et de la base de l'ouverture il sortit un sac vert, d'abord très petit, mais qui bientôt s'étendit de manière qu'il surpassa plusieurs fois la longeur du globule. Dans l'intérieur de ce sac parurent bientôt les spirales. Elles étaient accompagnées de leur points brillans, comme dans une conjugée entièrement développée. Le tube lui-même montrait ses cloisons, d'abord une, puis deux, puis un plus grand nombre; enfin la conjugée se détacha de son grain pour flotter seule sur le liquide; et alors, à la grandeur près, et aux deux extremités qui étaient encore pointues, elle ressembla parfaitement à la plante qui lui avait donné naissance.

pp. 69–70 Par rapport aux cloisons intérieures qui divisent le tube de la conjugée, elles sont ainsi que lui formées d'une membrane très-fine et transparente. Quoiqu'elles paraissent simples, j'ai lieu de les croire doubles; car j'ai souvent vu un tube de conferve se séparer en deux, en trois ou même en autant de parties qu'il tient de loges. Et comme ces loges, au lieu de se vider, retiennent chacune la matière verte ou les spirales, il est à présumer qu'elles étaient exactement fermées; car autrement cette apparence n'aurait pu avoir lieu.

On peut donc considérer les tubes des conferves dont il est ici question, non pas comme formant chacun une plante particulière mais plutôt comme l'assemblage d'un grand nombre de plantes. Sous ce point de vue chaque loge est elle même une plante qui ne communique point avec les autres renfermées dans le même tube. Elle peut leur être appliquée, elle peut aussi en être séparée: elle a son enveloppe particulière, ses spirales, ses grains, en un mot tout ce qui constitue plante, et comme nous le verrons bientôt, elle peut aussi se reproduire.

p. 70 1er jour. On aperçoit vers les grosses extremités de l'oeuf trois ou quatre globules assez gros (1/97 mill), ovalaires, d'abord séparés, mais qui au bout de quelques heures, se groupent irregulièrement. En les examinant avec attention, on voit qu'ils renferment dans leur intérieur d'autres globules beaucoup plus petits (globulins).

p. 70 2e jour. Le nombre des globules augmente, mais sans changer d'aspect. Ils forment une espèce de gâteau irrégulièrement festonné, un peu moins transparent au centre que sur les bords; on voit que là il y a déjà superposition de globules, tandis que sur ces derniers ils ne sont encore qu'accolés les uns aux autres.

p. 70 . . . nous savons que la propriété caracteristique des cellules, surtout à cette époque de la vie embryonnaire, est de servir en quelque sorte de matrice à d'autres cellules (globulins) qui se développent dans leur intérieur.

p. 71 Le globule embryonnaire s'est considérablement accru, et déjà il a doublé en grosseur . . . Le hile de son côté s'est prolongé et paraît formé de deux globules diaphanes, qui ne tardent pas à se séparer et à se détacher l'un de l'autre.

p. 71 Une notable métamorphose s'est déclarée dans le globule embryonnaire qui a pris une forme totalement différente de celle qu'il offrait hier. Sa périphérie s'est divisée en cinq lobes peu profonds; le centre du globule est plus diaphane que sa périphérie . . . et ce globule présente maintenant à sa surface des facettes irregulières.

p. 71 Alors s'opère un phenomène important: à l'intérieur des cellules primordiales, on commence à apercevoir des cellules secondaires qui, s'accroissant chaque jour de plus en plus, finissent par détruire les cellules primordiales dont les parois seules persistent, et deviennent un lacis de petits vaisseaux.

p. 71 Ainsi, c'est la surface du globule embryonnaire qui forme le premier tissu général, comme c'est la surface des grumeaux dont il se compose, qui devient le premier tissu cellulaire interne. Ainsi, la transformation originelle des fluides organisables en tissus s'opère par la solidification de leur surfaces.

p. 71 Nous avons vu dans le cours du développement embryonnaire deux modes de développement des tissus, celui du foie dont le tissu cellulaire s'augmente par des productions médianes, comme je l'ai indiqué le premier dans les végétaux, et celui du tissu dermo-musculaire qui se propage par l'accroissement centripète des canalicules qui forment le feutré d'infiltration que l'on remarque. Ceci renverse absolument l'uniformité de formation des tissus animaux, indiquée par Bordeaux, Meckel, etc., et l'on est forcé de reconnaître la pluralité de formation des tissus animaux admise par Bichat et son école.

p. 73 Die Verästelungen des Stammes entspringen immer an dem oberen Ende der Zellen des letzeren und sind von diesen Zellen durch eine Scheidewand getrennt . . . bildet sich an seiner Verbindungstelle mit der Stammstelle eine ringförmige ins Innere vorspringende Verengerung, welche den grünen Zelleninhalt an diese Stelle zusammenschnürt, also eine ringförmige in der Mitte durchbrochene Scheidewand darstellt.

p. 74 Je propose de nommer ainsi ce que d'autres observateurs ont appelé une gelée vivante, cette substance glutineuse diaphane, insoluble dans l'eau, se contractant en masses globuleuses, s'attachant aux aiguilles de dissection et se laissant étirer comme du mucus, enfin se trouvant dans tous les animaux inférieurs interposée aux autres élémens de structure.

p. 74 Immer bestehen sie aus einem granulösen *Parenchym*, dessen graurötliche, sehr kleine Körnchen von einem halbweichen, zähen, durchsichtigen, zellgewebeartigen Bindungstoffe durchzogen werden. In der Mitte oder in der Nähe derselben befindet sich ein runder oder länglich-runder *nucleus*, welcher aus einer begrenzenden Linie und einem ganz hellen Innern besteht.

pp. 74–5 Ich fand dieses Innere, zellenähnliche Gebilde welches ich aus später zu erörtenden Grunden *Primordialschlauch, utriculus primordialis* nennen will in gleich vollkommenen Zustande in einer Reihe von dicotylen Gewächsen. . . z.b. *Sambucus Ebulus, Ficus Carica, Pinus sylvestris*, etc. Das Vorgehende könnte schliessen lassen dass Hartig diesen Primordialschlauch gekannt und als *Ptychode* beschrieben habe.

p. 75 Da wie schon bemerkt diese zähe Flüssigkeit überall, wo Zellen entsehen sollen, den ersten, die künftigen Zellen andeutenden festen Bildungen vorausgeht, da wir ferner annehmen müssen, dass dieselbe das Material für die Bildung des Nucleus und des Primordialschlauches liefert, in dem diese nicht nur in der nächsten räumlichen Verbindung mit derselben stehen, sondern auch auf Jod auf analoge Weise reagiren dass also ihre Organisation der Process ist, welcher die Entstehung der neuen Zellen einleitet, so mag es wohl gerechtfertigt sein, wenn ich zur Bezeichnung dieser Substanz eine auf physiologische Function sich beziehende Benennung in dem Worte *Protoplasma* vorschläge.

Chapter 8

p. 81 . . . gelblich schimmernde, dunkle Fleck . . . einmal sah ich statt eines einzigen Flecks zwei kleinere, dicht beisammen leigende.

p. 81 Ich bin auf diesen Fleck aufmerksam geworden, weil ich denselben auch bei anderen Thierklassen begegnete; ob bei Wirbelthieren constant, bin ich noch zweifelhaft; sehr deutlich aber fur jeden Beobachter ist dieser Fleck bei *Phalangium opilio*.

p. 81 Diesen Fleck, den ich wenigstens bei Säugethieren für constant halten möchte, nenne ich den Keimfleck (macula germinativa).

p. 81 Der Keim ist bei seinem ersten Auftreten eben das, was ich Keimfleck genannt habe . . . habe ich deutlich die Entstehung der Keimschicht aus dem Keimfleck beobachtet.

Chapter 9

p. 84 Conspicitur tunc in interna eius facie vesicula prominula diaphana, halone exili materiei albae globosae, quae in ovulo maturo cumulum constituit, circumdata . . . Consistentia nempe vesiculae admodum tenera est, ut in minoribus ovulis ad instar bullae acqueae a contrectatione levissima dissiliat.

p. 85 Primum quidem, si cicatriculum inquiras, postquam vitellus ab infundibulo iam exceptus est, vesiculam quae prius in cicatricula ovuli descripta fuit, nusquam deprehendes.

p. 88 . . . schöner, fast wie Pflanzenzellgewebe aussehender sechseitiger Balken und in welchen kleine Körnchen von runder Form . . . sich befinden.

p. 88 Jeder dieser Kugel hat überall eine äussere mehr oder minder deutliche zellgewebige Hülle und enthält eine eigene Parenchymmasse, einen selbstständigen Nucleus oder Kern, und einen in diesem enthaltenen rundlichen, durchsichtigen zweiten Nucleus.

p. 89 Es besteht aus dicht neben einander liegenden, rhomboidal oder quadratisch rundlichen Zellen deren Begrenzungen von einfachen, fadenartigen Linien gebildet werden – in jeder Zelle ohne Ausnahmen befindet sich ein etwas dunklerer und compakter Nucleus von runder oder länglich runder Form. Er nimmt grösstentheils die Mitte der Zelle ein, besteht aus einem feinkörnigen Wesen, enthält aber in seinem Inneren ein genau rundes Körperchen, welches auf diese Weise in ihm selbst wiederum eine Art von zweitem Nucleus bildet.

p. 89 . . . enthält aber in ihrer Mitte im Inneren einen dunkelen runden Kern, eine Formation, welche an den im Pflanzenreiche vorkommenden Nucleus erinnert.

p. 92 Somit reducirt sich der thierische Organismus fast ganz in drei Elementar-Hauptformen: die flüssige, die körnige und die faserige.

p. 92 Die körnige Grundform dringt wieder eine Analogie mit der Pflanze auf, welche bekanntlich beinahe ganz aus Körnern oder Zellen zusammengesetzt ist.

p. 93 In Bezug auf die Bedeutung der gangliösen Körperchen wäre zu bemerken, dass sie wahrscheinlich Centralgebilde sind, wofür ihre ganze dreifach concentrische Organization spricht, und die sich zu den elementaren Hirn – und Nervenfasern wie Kraftcentra zu Kraftleitungslinien, wie Ganglien zu Gangliennerven, wie die Hirnmassen zum Rückenmark und Hirnnerven sich verhalten möchten. Sie wären Sammler, Erzeuger und Vertheiler des Nervenorgans.

CHAPTER 10

p. 96 Solche Vergleiche hatten aber desshalb keine weitere Folge, weil es bloss einzelne Formähnlichkeiten von Gebilden waren, welche die manchfaltigsten Formen zeigen.

p. 96 Von diesem Moment ab richteten sich alle meine Bemühungen darauf den Beweis für die Präexistenz des Kernes vor den Zellen zu erbringen.

p. 97 An beiden Orten entstehen nun im Gummi sehr bald die oben erwähnten kleinen Schleimkörnchen, wodurch die bis dahin homogene Gummilösung sich trübt, oder bei grösserer Menge von Granulis selbst opak wird. Darauf zeigen sich in dieser Masse einzelne, grössere schärfer gezeichnete Kernchen, und bald nachher treten auch die Cytoblasten auf, die gleichsam als granulöse Coagulationen, um jene Kernchen erscheinen. – Die Cytoblasten wachsen aber in diesem freien Zustande noch bedeutend.

p. 98 Sobald die Cytoblasten ihre völlige Grösse erreicht haben erhebt sich auf ihnen ein feines, durchsichtiges Bläschen, dies ist die junge Zelle, die anfangs ein sehr flaches Kugelsegment darstellt, dessen plane Seite vom Cytoblasten, dessen Convex-Seite von der jungen Zelle gebildet wird, die auf ihm etwa wie ein Uhrglas auf einer Uhr aufsitzt.

p. 98 Nach und nach wächst nun die ganze Zelle über den Rand des Cytoblasten hinaus, und wird rasch so gross, dass endlich der letzere nur als ein kleiner in einer der Seitenwände eingeschlossener Körper erscheint. . . . Noch immer findet man den Cytoblasten in der Zellwandung eingeschlossen, an welcher Stelle er den ganzen Lebensprocess der von ihm gebildeten Zelle mit durchmacht, wenn er nicht bei Zellen, die zu höherer Entwickelung bestimmt sind, entweder an seinem Ort, oder, nachdem er gleichsam als unnützes Glied abgestossen ist, in der Höhlung der Zelle aufgelöst und resorbirt wird.

p. 102 Die vorliegende Abhandlung hat zur Aufgabe den inningsten Zusammenhang beider Reiche der organischen Natur aus der Gleichheit der Entwicklungsgesetze der Elementartheile der Thiere und Pflanzen nachzuweisen. Das Hauptresultat der Untersuchung ist, dass ein gemeinsames Entwicklungsprinzip allen einzelnen Elementartheilen aller Organismen zum Grunde liegt, ungefähr so wie alle Krystalle trotz der Verschiedenheit ihrer Form sich doch nach denselben Gesetzen bilden.

p. 102 Die Theile einer animalischen Mischung, wenn sie aus dem flüssigen Zustand in den festen übergeht, vereinigen sich in kleine Fasern und Blättchen. Diese Veränderung geschieht durch die Mischung eigene, und durch äussere Umstände modificirte Central- und Cohäsionskraft, und dann kann als eine animalische Krystallisation angesehen werden.

p. 104 Zu einer durchgeführten Vergleichung dieser Gebilde gab die Pariser Preisfrage über denselben Gegenstand Veranlassung.

p. 104 Die Construktionen der Muskel-und Nervenfasern erwarten zu richtiger Leitung der Theorie noch viel mehr empirische Data als bisher vorhanden sind.

p. 104 Es scheint, wie wenn er sich durch Schleidens glückliche Untersuchungen über Genesis des Pflanzengewebes zu sehr hinreissen lassen, und dass er die Analogien aus dem Pflanzenreich auf die Thierorganisation in zu grosser Allgemeinheit übertragen.

p. 104 Uns hat Schwanns Buch grössenteils den Eindruck ächten naturwissenschaftlichen Geistes zurückgelassen.

p. 104 Uebrigens behalten die ausgesprochenen theoretischen Grundsätze, bei der so offenbaren von Schwann selbst bemerkten, Relativität der Zelle und des gekernten Körnchens in allgemeinere Sinne ihre Giltigkeit, womit sowohl die Idee als das Verdienst geistreicher Entwicklung und reicher Zusammenstellung des empirischen Materials dem Autor als unantastbares Gut gewonnen bleibt.

Chapter 11

p. 106 . . . die physiologische Grundlage und Auffassung konnte ihr aber erst zu Theil werden, nachdem Schleiden und Schwann die grossartigen Entdeckungen von dem gemeinschaftlichen Zellleben der höheren Organismen gemacht haben.

p. 106 Schwann hat uns in seinem schätzbaren Werke auch über diesem Gegenstand vieles wichtige mitgetheilt.

p. 107 Encyclopädisches Wörterbuch der medizinischen Wissenschaften.

p. 107 . . . enthalten einen runden Kern . . . in dem sich meist wieder ein Nucleus erkennen lässt.

p. 108 In den meisten pflanzlichen und thierischen Geweben kommen während des ganzen Lebens oder zu einer gewissen Zeit ihrer Entwickelung mikroskopische Körperchen von eigenthümlicher und sehr characteristischer Form vor, welche man mit den oben angeführten Namen zu bezeichnen pflegt. Es sind Bläschen bestehend aus einer feinen Haut und einem flüssigen, mitunter etwas körnigen Inhalte; in ihrer Wand liegt ein kleinerer, dunklerer Körper, der Zellenkern, *Nucleus*, cytoblast (Schleiden), und dieser ist in der Regel ausgezeichnet durch einen or zwei, selten mehr noch dunklere und fast regelmässig runde Fleckchen, *Nucleoli*, Kernkörperchen.

p. 108 Andere Beobachtungen machen es zweifelhaft, ob die granulöse Substanz, aus welcher der Zellkern hervorgeht, sich nur in der Umgebung eines Kernkörperchens niederschlagen könne.

p. 108 Fürs erste giebt es, wie Kerne ohne Kernkörperchen, so auch Zellen ohne Kern.

p. 108 Dass Zeugung von Zellen in Zellen auch im thierischen Organismus vorkomme, ist nicht mehr zweifelhaft. . . .

p. 110 Die Bewegungen waren nicht oscillirend, sondern bald fortschreitend, bald rückschreitend, bald Seiten – bald wälzende Bewegungen. Einzelne Körperchen wiechen einander aus und näherten sich einander, und im letzeren Falle wurden beider Bewegungen lebhafter, tiefe schwimmende tauchten auf, oberflächliche senkten sich nieder und so glich dieses wunderbare, allerdings Staunen erregende Schauspiel einem Heere von Monaden voll innerer Lebendigkeit, voll innerer in Bewegungen sich offenbarender Selbstbestimmung.

p. 110 Diese wenigen Beobachtungen und Versuche mögen vor der Hand hinreichen, einen Fingerzeig zu geben, dass uns die mikroskopische Molecularwelt ungeachtet der Forschungen die besonders in der neuen Zeit so viele glänzende Resultate lieferten, noch keineswegs hinlänglich bekannt ist, und dass daher nichts dringender zu wünschen wäre, als dass Hr. Prof. Ettinghausen, im Besitze eines so vortrefflichen Instrumentes, den genannten Gegenstand ja gewiss mit gehöriger Musse weiter verfolgen möchte.

p. 111 Dasjenige, was die Primordialzelle am bestimmtesten charakterisirt und ihre Bedeuting im Leben der Pflanzen im Allgemeinen, nämentlich aber der Schwärmzellen als das wesentliche Moment erscheint, ist, dass sie das Contractile am Pflanzenorganismus ist, das heisst, dass sie die Fähigkeit besitzt, in Folge innerer Thätigkeit ihre Gestalt, ohne entsprechende Veränderung ihres Volumens zu verändern."

p. 112 Die zweite Frage, auf welche Weise die Entstehung neuer Zellen in einem schon ausgebildeten Zellgewebe vor sich geht, hängt nothwendig mit der Ansicht zusammen, wie man überhaupt die Bildung der Zellen vorstellt.

p. 112 Ich meine Theils habe mich nie zu der Ansicht bekannt, dass die Cytoblasten die Quelle neuer Zellen in der Art seien, dass dieselben unmittelbar von ihnen in ihrer räumlichen Ausbildung ausgehen, und besonders in dem gegebenen Falle würde es sehr schwer halten, die Bildung neuer Zellen in solchen Internodien zu erklären, deren Zellen meist ohne Zellenkern sind. Doch mein Hauptargument gegen diese Theorie ist das, dass man das Hervortreten der jungen Zellbläschen aus dem Zellkern nicht beobachtet, wenigstens dort nicht, wo Neubildungen stattfinden.

p. 113 Achtet man auf das Vorkommen solcher zarten Zellwände noch etwas genauer, so wird man nicht übersehen können, dass dieselben sich meist wie Querwände in nach irgend eine Richtung sich ausdehnenden Zellen verhalten, wodurch dieselben gleichsam in zwei Fächer getheilt werden.

p. 113 . . . in den meisten Fällen, wo Wachstum der Zellgewebsmassen erfolgt, dieselbe nicht durch intrautriculare, sondern durch meristematische Zellbildung erfolgt, daher auch weder von Mutterzellen noch von ihrer Auflösung die Rede sein kann.

p. 114 Ich schliesse daraus, dass in den Conferven die Vermehrung der Zellen unmöglich durch Bildung kleiner freier Zellen in Innern geschehen kann.

pp. 114–5 . . . dass bei diesen Abtheilungen des Pflanzenreiches mit Ausnahme der Specialmutterzellen bloss die freie Zellenbildung um einen Kern vorkomme.

p. 115 Zuweilen bildet sich in einer Zelle eine Keimzelle bloss aus ihrem eigenen Inhalte.

p. 115 Freie Zellenbildung mit sichtbaren Kern habe ich bis jetzt mit Sicherheit im Embryosack der Phanerogamen beobachtet.

p. 115 Es entsteht im Inhalte der Mutterzelle ein Kern. Derselbe sammelt, durch Attraction, an seiner Oberfläche eine grössere oder geringere Menge Inhaltes der Mutterzelle, welcher, an seinem Umfange wenigstens, aus homogenem Schleim besteht. Diese Inhaltspartie bekleidet sich an ihrer ganzen Fläche mit einer Membran.

Chapter 12

p. 117 . . . qua propter vias quas Ova transire debent, iterum iterumque perquisivimus, ac invenimus in Oviductus dextro medio unum, in eiusdem lateris cornu extremo duo minutissima Ova qualia exhibet.

pp. 118–9 Cette ligne qui, d'abord, ne se dessinait à la surface de l'oeuf que par une très légère dépression, se creuse avec une inconcevable rapidité, et détermine la formation d'un nombre considérable de petites rides parallèles entre elles et perpendiculaires à sa propre direction, qui prennent naissance dans le sillon qu'elle produit. Celui-ci devient toujours plus profond, et l'oeuf se trouve bientôt divisé en deux segmens très-prononcés. Il se manifeste alors une nouvelle ligne, mais celle-ci passe a peu près sur la limite qui sépare les deux hémisphères brun et jaune, et coupe l'oeuf circulairement comme une espèce d'équateur. L'hémisphère brun était partagé en quatre portions égales, chacune d'elles se divise en deux au moyen de nouvelles dépressions parallèles au sillon qui s'était montré le premier. Cet hémisphère se trouve alors divisé en seize parties égales ou à peu près. Dès lors la partie brune de l'oeuf se trouve divisée en un certain nombre de granulations analogues à celles d'une framboise. On en compte d'abord trente ou quarante, mais au bout de deux heures elles se sont elle-mêmes sous-divisées, et leur nombre s'élève à plus de quatre-vingts.

p. 120 Si l'on fait prendre au germe un certain degré de consistance, au moyen de l'ébulition, ou par quelqu'autre procédé, et si l'on saisit pour faire cette opération l'époque, pendant laquelle sa surface, sur-tout la brune, est toute sillonée en divers sens, on peut après séparer le germe en plusieurs masses qui sont plus ou moins grandes, selon que les sillons à la surface étoient plus ou moins multipliés; en un mot, nous trouvons, en répétant cette expérience à diverses époques, que toute la matière qui constitue le germe, se divise d'abord en deux, puis en quatre parties, lesquelles se divisent et se subdivisent en d'autres plus petites.

p. 120 . . . ihre Beobachtungen über die auffallenden Furchungen die auf der Oberfläche der Froscheier zu erkennen sind.

p. 120 . . . ohne Zweifel weil ihnen kein Mittel bekannt war, das Eiweiss zu entfernen, um die Dotterkugel zu erhärten um sie Zergliederung zu unterwerfen.

p. 120 Wir werden dabei immer von dem an der Oberfläche Sichtbaren beginnen und dann zu dem Inneren fortschreiten, wollen aber zum besseren Verständniss gleich von vorn hinein bemerken, dass die an der Oberfläche sichtbaren Spalten nichts sind als die Gränzen von Theilungen, die die ganze Dotterkugel erleidet.

pp. 121–2 Unter allen Eiern, die ich kennen gelernt habe, scheinen mir dagegen die Froscheier als die einzigen an denen die Präexistenz widerlegt werden kann.

p. 124 Doch mag dies hinreichen um zu zeigen, dass solche aus Dotterkörnchen bestehenden und mit einem eigenthümlichen hellen Fleck versehenen Klumpen aus welchen nun der ganze Dotter besteht, Theile sind, deren Identität mit den Zellen aus welchen der Embryo sich nun alsbald aufbaut, wegen des Uebereinstimmens in Bezug auf diese characteristischen Bestandtheile als unzweifelhaft angesehen werden darf.

p. 125 Zerdrückt man jetzt die Dotter, so findet man, dass ein jeder solcher Hügel vor dem kleinern Theile einer rundlichen Zelle dargestellt wird, und dass überhaupt der ganze Dotter, also auch sein mittlerer Theil, aus lauter Zellen besteht.

p. 126 In der Mitte befinden sich grössere Zellen ohne Kern. Sie sind in Bezug auf die zu erzeugende junge Generation die am meisten zurückstehenden; sie befinden sich als Mutterzellen auf der Lebenstufe, wo der Zellkern resorbirt ist, und auf Kosten des Zelleninhaltes die junge Brut sich entwickeln soll.

p. 126 Der Furchungsprocess der Batrachier Dotter ist nämlich nichts anderes, als ein allmählig fortschreitender Geburtsact vielfach eingeschachtelter Mutterzellen ist, welche zum Aufbau des Gesammt-Zellen-Organismus dienen sollen.

p. 126 Alle diese Momente betreffen aber, wie aus den obigen Untersuchungen sich ergeben hat, nicht den Bildungsprocess der Furchungskugelzellen, sondern vielmehr das Freiwerden, die Geburt der bereits gebildeten, noch kernlosen Brutzellen aus ihren Mutterzellenmembranen.

Chapter 13

p. 130 Mir selbst war die extracellulare Entstehung thierischer Zellen, seit dem Bekanntwerden der Zellentheorie, ebenso unwahrscheinlich, wie die Generatio aequivoca der Organismen.

p. 130 An dieser Stelle ist es nur mein Zweck, vorläufig die Aufmerksamkeit auf das Ergebniss zu lenken, dass sich in den Anlagen der verschiedenartigsten Gewebe fortschreitende Theilung vorhandener Zellen, nirgends aber das Auftreten extracellularer Kerne oder extracellularer Zellen beobachten lässt.

p. 131 Diese Ergebnisse haben zur Pathologie eine ebenso nahe Bezeihung wie zur Physiologie. Es kann kaum noch bestritten werden, dass die pathologischen Gewebeformen nur Varianten der normalen embryonischen Entwickelungstypen bilden und es ist nicht wahrscheinlich dass sie das Vorrecht der extracellularen Entstehung von Zellen besitzen sollen. Die sogenannte 'Organisation der plastischen Exsudate' und die früheste Bildungsgeschichte der krankhaften Geschwülste bedarf in dieser Hinsicht einer Prüfung. Gestützt auf die Bestätigung, welche meine vieljährigen Zweifel erfahren, wage ich die Vermutung auszusprechen, dass die pathologischen Gewebe ebensowenig wie die normalen in einem extracellularen Cytoblastem sich bilden, sondern Abkömmlinge oder Erzeugnisse normaler Gewebe des Organismus sind.

p. 132 Es ist kaum nöthig, die Frage nach der Aehnlichkeit oder Verschiedenheit von Zellen und Krystallen einer besonderen Erörterung zu unterwerfen, da beide Gebilde nach den vorliegenden Thatsachen keine Vergleichungspunkte bieten.

p. 133 Nach der Ansicht Schwanns war die Intercellularsubstanz Cytoblastem, für die Entwickelung neuer Zellen bestimmt. Dies halte ich für unrichtig.

p. 133 Auch in der Pathologie können wir soweit gehen, als allgemeines Princip hinzustellen, dass überhaupt keine Entwickelung *de novo* beginnt, dass wir also auch in der Entwickelungsgeschichte der einzelnen Theile, gerade wie in der Entwickelung ganzer Organismen die Generatio aequivoca zurückweisen.

p. 133 Nirgends gibt es eine Art der Neubildung, als die fissipare; ein Element nach dem andern theilt sich; Generation geht aus Generation hervor.

p. 134 Vielleicht ist es in heutiger Zeit ein Verdienst, das historische Recht anzuerkennen, denn es ist in der That erstaunlich, mit welchem Leichtsinn gerade die jenigen, welche jede Kleinigkeit, die sie gefunden haben, als eine Entdeckung preisen, über die Vorfahren aburtheilen.

p. 135 Eine solche Vertheidigung ist keine That eitlen Ehrgeizes, kein Aufgeben des rein wissenschaftlichen Strebens. Denn wenn wir der Wissenschaft dienen wollen, so müssen wir sie auch ausbreiten, nicht bloss in unserem eigenen Wissen, sondern auch in der Schätzung der Anderen.

p. 135 In einer so unmittelbar praktischen Wissenschaft, wie die Medicin, in einer Zeit so schnellen Wachsens der Erfahrungen, wie die unsrige, haben wir doppelt die Verpflichtung, unsere Kenntniss der Gesammtheit der Fachgenossen zugänglich zu machen.

p. 136 Für die normalen Gewebe ist bisher keine sichere Ausnahme bekannt. Aber unter pathologischen Verhältnissen ist die Entstehung von Zellen innerhalb von Zellen, auf endogene Wege, nach den Beobachtungen von Hiss, Buhl, Weber und mir, unzweifelhaft.

p. 136 Es wäre eine der interessantesten Entdeckungen, wenn es gelänge nachzuweisen, dass nicht bloss in krankhaften Zuständen, welche die Zerstörung der normalen Gewebe herbeiführen, sondern auch bei der normalen Entwickelung oder beim Wiedersatz zerstörter Gewebe neben der Zelltheilung auch endogene Zellen-und Kernbildung innerhalb von Zellen oder in Aequivalenten von Zellen vorkommt.

Chapter 14

p. 139 Ueber die Art und Weise der Kerntheilung sind die Beobachtungen keineswegs so weit gehend wie die entspechenden über das Verhalten der Zellmembranen.

p. 139 Die Regel ist, dass das Kernkörperchen sich in zwei Theile abschürt, und ebenso der Kern in zwei Kerne.

p. 139 Nun erst theilt sich dieser Kern in zwei runde Kerne, und dann geht die Bildung der zwei secundären Specialmutterzellen vor sich.

p. 140 Die gewöhnlichste Art der Kerntheilung geht in der Weise vor sich, dass zuerst an einer Seite des gewöhnlich etwas ovalen Kernes eine kleine Einschnürung oder Einkerbung sich bildet, die sich nach und nach über die Fläche des Kernes herüber erstreckt und von der aus sich dann die Scheidewand durch das Innere des Kernes hindurch schiebt. Zuweilen sieht man auch gleichzeitig an zwei oder an mehren Stellen des Kernumfanges solche Einkerbungen und nachher dem entsprechend eine zwei- oder mehrfachen Theilung des Kernes. Immer ist jedoch die Scheidewand zunächst vollkommen gerade und erst in dem Masse, als die Kerntheile wachsen und sich zu besonderen, auch äusserlich getrennten Kernen umbilden, rundet sich auch ihre Begrenzung ab und sie entfernen sich endlich von einander.

p. 140 Wie er sich bei der Theilung verhält, ist mir gerade für den wichtigsten Moment entgangen, doch sei bemerkt, dass ich öfters ein Stadium sah, in welchem die Kerne aussnehmend lang gezogen, manche auch mit Einschnürungen versehen waren, so dass ich, auch ohne einen getheilten Kern gesehen zu haben, doch auf Theilung schliessen darf, wozu noch kommt, dass in keinem Falle die Zelle des Kernes entbehrte.

p. 141 . . . die Vermehrung der Kerne, welche die einzige Art ist, wie die Kerne für die in den primären Zellen sich bildenden Generationen endogener Zellen entstehen, sah ich in der Weise, dass dieselben länger wurden, in der Mitte sich einschnürten und endlich in zweie zerfielen.

p. 141 Was aber die hellen bläschenförmigen Kerne innerhalb der Furchungskugeln betrifft, so glaube ich zu dem Resultate gekommen zu sein, dass sie bei jeder bevorstehenden neuen Theilung schwinden, und erst nach dem diese zu Ende gebracht worden, wieder neugebildet zum Vorschein kommen.

p. 145 Etwas später werden beide, der Pronucleolus und Pronucleus auf ganz eigenthümliche Weise gestreift; auf ihrer Oberfläche bemerkt man eine Menge schlangenförmige aus dichterer und glänzenderer plasmatischer Substanz bestehender Linien, die als Differenzirungslinien der beiden Pronuclei sich darstellen. In den folgenden Staden werden diese Linien in breite sehr dichte und glänzende Leisten umgewandelt, welche als vorspringende Meridiane sich darstellen und auch nur durch Einwirkung des Wassers zur Anschaung kommen. Die Theilungslamelle des Pronucleus ist auch erweitert, sie nimmt eine aequatoriale Stellung ein und ist aus dichteren, glänzenden plasmatischen Klumpen aufgebaut. Um diese Zeit bemerkt man an den zwei Polen, wo die Meridianleisten convergiren, zwei plasmatische, zuerst weniger dichte Regionen, welche später entstehen und zwei künftigen secundären Nuclei darstellen.

p. 146 . . . von 5 Tage an sieht man nämlich oft zwischen gewöhnlichen Endothelien Zellen, die in ihrem Kerne einige oder sehr viele längliche glänzende stab- oder fadenförmige Körper enthalten, welche letzere häufig stark gewunden und geschlängelt sind, ja auch förmliche Fadenknäuel bilden.

Chapter 15

pp. 150–1 Bei dieser Contractilität des Protoplasma sind Gestaltveränderungen der ganzen Zelle durch Anwesenheit einer starren Zellenmembran natürlich verhindert oder ganz unmöglich gemacht. Je weniger vollkommen aber die Oberfläche des Protoplasma zu einer Membran erhärtet ist, je näher die Zelle dem ursprünglichen membranlosen Zustande sich befindet, auf welchem sie nur ein nacktes Protoplasmaklümpchen mit Kern darstellt, um so freier und ungehinderter können sich die Bewegungen äussern.

p. 151 . . . so kommt bei den Radiolarien die Bewegungsfähigkeit ausschliesslich dem extracapsularen Sarkodekörper zu.

p. 152 . . . dieser weist also die Grösse der Anziehung eines Moleküls des fraglichen Körpers in verdünnter wasseriger Lösung zum Wasser an.

p. 152 Wird eine ausgewachsene Zelle in eine starke Salzlösung gebracht, so löst sich bekanntlich der lebendige Plasmaschlauch von der Zellhaut los, und zeiht sich auf ein kleineres Volumen zusammen, indem der von ihm umgeschlossene Zellsaft Wasser an die umgebende Salzlösung abgiebt. Je schwacher die eindringende Lösung, um so geringer ist diese Contraction oder die Plasmolyse.

Chapter 16

p. 155 Gleich nach dem Erscheinen dieser Form werden die Fäden in der Mitte der Tonne etwas dicker (die Verdickungen liegen gewöhnlich nicht in einer Ebene); dann reissen die Fäden in diesen Verdickungen und die Tonne teilt sich in zwei gleiche Teile welche sich zugleich von einander entfernen. Dadurch sind die zwei neuen Kerne gebildet.

p. 157 Zweitem lässt sich hier feststellen, dass eine Längsspaltung der Fäden in der Aequatorialebene, welche Strasburger nach seinem obigen Wortlauf dann hier zu Hülfe nehmen müsste, ebenfalls für diesen Zweck nicht zu verwerthen ist: denn wie ich ausführlich beschrieben habe beginnt die von mir gefundene Fädenlängsspaltung bei Salamander bereits in den Knäuelformen . . . und dauert durch die Sternformen.

p. 158 Nachdem das Ei eine Zeitlang in diesem Zustand verharrt hat, erscheint, wie ich stets beobachtete, ein helles Bläschen an der Vagina zugewendeten Eipol und nach einiger Zeit ein zweites derartiges Bläschen in geringer Entfernung von dem ersteren . . . Natürlicherweise ist die Entstehung dieser Bläschen nicht direkt sichtbar, sondern sie markiren sich erst, wenn sie eine gewisse Grösse erreicht haben so dass die Möglichkeit, dass das zweite Bläschen ein Abkömmling des ersten sei, wenn auch für unwahrscheinlich so doch nicht für völlig wiederlegt erachten muss.

p. 159 La pénétration a lieu en un point quelconque de la surface du vitellus. Je suis d'avis que la fécondation normale de l'Etoile de mer se fait à l'aide d'un seul zoosperme par oeuf; chez l'Oursin, ce fait est tout à fait évident.

p. 159 Le point de pénétration devient le centre d'une étoile ou aster mâle; dans le milieu de l'aster se forme un amas ou pronucléus mâle qui va se fusionner avec le pronucléus femelle d'une manière tout à fait conforme á ce qui s'observe chez l'Oursin.

p. 162 (1) Que non seulement le noyau chromatique du zoosperme, mais aussi la couche achromatique qui l'entoure (couche périnucléaire) interviennent dans la formation du pronucléus mâle. (2) Que la vésicule germinative fournit au pronucléus femelle non seulement des éléments chromatiques, mais aussi un corps achromatique. (3) Que les deux pronucléus, sans se confondre, peuvent acquérir, par le fait de leur maturation progressive, la constitution de noyaux ordinaires. (4) Que chez l'ascaride du cheval il ne se produit pas un noyau unique aux dépens des deux pronucléus; qu'il n'existe pas un 'Furchungskern' dans le sens que O. Hertwig a attaché à ce mot. L'essence de la fécondation ne réside donc pas dans la conjugaison de deux éléments nucléaires, mais dans la formation de ces éléments dans le gonocyte femelle. L'un de ces noyaux dérive de l'oeuf, l'autre du zoosperme. Les éléments nucléaires expulsés sous forme de globules polaires sont remplacés par le pronucléus mâle et dès que deux demi-noyaux, l'un mâle, l'autre femelle se sont constitués, la fécondation est accomplie. (5) A la suite d'une série de transformations que subit la charpente nucléaire de chacun des pronucléus, transformations identiques d'ailleurs à celles qui se produisent dans un noyau en voie de division, chaque pronucléus donne naissance à deux anses chromatiques. (6) Les quatres anses chromatiques interviennent dans la formation de l'étoile chromatique (plaque nucléaire); mais elle restent distinctes. Chacune d'elles se divise longitudinalement en deux anses secondaires jumelles. (7) Les noyaux des deux premiers blastomères reçoivent chacun une moitié de chaque anse primaire, soit quatre anses secondaires, dont deux mâles et deux femelles.

Il ne se produit donc de fusions entre la chromatine mâle et la chromatine femelle à aucun stade de la division . . .

Les éléments d'origine mâle et femelle ne se confondent pas en un noyau de segmentation et peut-être restent-ils distincts dans tous les noyaux derivés.

p. 164 . . . die einleitenden Phasen der Kerntheilung, die ich die Prophasen nennen will, beginnen mit der Ausbildung des Fadenknäuels.

p. 164 Ich werde, im Gegensatz zu den Prophasen die mit der Längspaltung der Segmente in der Kernplatte ablaufen, die Stadien von beginnenden Auseinanderweichen der Tochtersegmente bis zur vollendeten Trennung und Umlagerung darstellend, als Metaphasen zusammenfassen.

p. 164 Die Phasen, die von der vollendeten Sonderung der Tochtersegmente . . . bis zur Fertigstellung der Tochterkerne verlaufen, können als Anaphasen der Theilung zusammengefasst werden.

p. 165 Unter dem Namen Telokinesis beschreibe ich gewisse Bewegungen des Kerns und des Mikrocentrums, welche gegen das Ende der Mitose hin stattfinden, und in so ferne wenigstens völlig typisch ablaufen, als sie immer einen ganz bestimmten Schlusseffekt zu Stande kommen lassen.

p. 165 Bei einer detaillierten Analyse des sog. Ruhekerns hat es sich gezeigt, dass die Konfiguration des Caryotins verschieden ist, wenn der Kern soeben in den Ruhestand getreten oder wenn er lange darin geblieben ist. Es schien daher geeignet, den Zwischenstand zwischen zwei aufeinanderfolgenden Teilungen mit einem besonderen Namen zu belegen. Ich bezeichne ihn wie eine Interphase, und spreche erst bei langer Interphase von einer typischen Ruhe des Kerns.

Chapter 17

p. 166 Die kernlosen Elementarorganismen sind es, welche wir als 'Cytoden' bestimmt von den echten (kernhaltigen) Zellen unterscheiden müssen.

p. 166 Die Cytoden oder die kernlosen Plasmaklumpen zerfallen gleich den echten kernhaltigen Zellen in zwei Gruppen.

pp. 171–2 Diese beiden 'haploiden' Kerne werden bei der Befruchtung zu einem 'diploiden' addiert, der nun 2a, 2b usw. enthält; und diese Doppelserie erbt sich durch die Spaltung eines jeden Chromosoma und durch die geregelte karyokinetische Vertheilung der Tochterchromosomen zunächst auf die beiden primären Furchungszellen fort. In den entstehenden ruhenden Kernen gehen die einzelnen Chromosomen scheinbar unter. Allein wir besitzen die wichtigsten Indizien für die Annahme, dass im Gerüst des Ruhekerns jedes in den Kern eingegangene Chromosoma als ein bestimmter Bezirk fortbesteht, um bei der Vorbereitung zur nächsten Teilung als das gleiche Chromosoma wieder zu erscheinen (*Theorie der Individualität der Chromosomen*). So erben sich also die bei der Befruchtung zusammengeführten zwei Chromosomenserien auf alle Zellen des neuen Individuums fort. Nur in den Fortpflanzungszellen wird durch die sog. Reduktionsteilung die doppelte Serie wieder auf die einfache herabgesetzt. Aus dem diploiden Zustand entsteht wieder der haploide.

Index